Gertrude Hoyt Memorial

M. Biery

COMPETITIVE
MANUFACTURING

COMPETITIVE MANUFACTURING

Using Production as a Management Tool

Stanley S. Miller, D.C.S.

 VAN NOSTRAND REINHOLD COMPANY
New York

98119

Printed in the United States of America
Designed by Azuretec Graphics

Van Nostrand Reinhold Company Inc.
115 Fifth Avenue
New York, New York 10003

Van Nostrand Reinhold Company Limited
Molly Millars Lane
Wokingham, Berkshire RG11 2PY, England

Van Nostrand Reinhold
480 La Trobe Street
Melbourne, Victoria 3000, Australia

Macmillan of Canada
Division of Canada Publishing Corporation
164 Commander Boulevard
Agincourt, Ontario M1S 3C7, Canada

16 15 14 13 12 11 10 9 8 7 6 5 4 3 2

Library of Congress Cataloging-in-Publication Data
Miller, Stanley S.
    Competitive manufacturing.
    Bibliography: p.
    Includes index.
    1. Strategic planning.   2. Production management.
I. Title.
HD30.28.M52   1987   658.5'03   87-10484
ISBN 0-442-26395-3

Contents

Preface

A major restructuring of American industry is taking place in the 1980s: a consolidation and modernization of facilities and a rationalization of the overall manufacturing processes they support. In this vast and hectic changeover, new tools and concepts have emerged and a new kind of industrial leadership has come to the fore. Manufacturing plants have closed, but the survivors are more viable in world competition.

The transformations of some corporations are the opposite of the diversification and acquisition strategies of the 1950s and 1960s, when companies were trying to branch out and cover more ground. The objective now is to return to fundamentals—strengthen the work disciplines, untap constructive views from all levels of the organization, concentrate, and do a superior job in selected fields of competition.

In other corporations, however, the diversification and acquisition movement has not run its course. Bringing order into the increasing proliferation of business commitments will continue to pose a major challenge for the manufacturing people.

Manufacturing personnel must play an active role in this restructuring, but they cannot work in isolation. We are not talking about improvements in productivity that can be left to the operating levels of the factories. We are talking about a complete shake-out in management thinking in order to enable manufacturing considerations to become an integral part of the planning and strategizing, the decision making, and the setting of priorities at the top levels of the organization. We are talking about bringing manufacturing capabilities into the fabric of the competitive action plans of every strategic business unit in the corporation. This is not a trivial exercise, certainly not after decades when manufacturing was considered a kind of service function set up to carry out the strategies devised elsewhere by more senior people.

The purpose of this book is to help general managers incorporate manufacturing considerations into their planning, and to help all the functional organizations work with each other to implement the business strategies. The Management Row Game provides a conceptual framework for this task and the workshop process provides a tool to apply it. They will help to identify interfunctional problems, work the interfaces, and shake down a strategy while setting into motion the action plans to carry it out. Like any other game, the Management Row Game is part toy, part intellectual exercise, and part reality; like any other game, it should provide a certain amount of satisfaction to the players.

I wish you joy of it, and hope it will help take you where we will all need to go if we are to give our industrial economy any real sense of purpose for the 1990s.

Acknowledgments

Many people have shared the experiences described in this book with me and have had an important role in shaping the outcome. I want to thank William L. Wallace and Milton J. Gottschalk, who were with me so much of the way. I would also like to thank the people who made these ventures possible by putting their own resources or their own time at stake: George Beck and Robert Packard, John Mahoney and Ronald Adams, Lawrence Wilson and John Mulroney, Chuck Prizer, Jack Weaver, Roy Djuvik, and John Fryling, among others.

I want to express my appreciation for the many stimulating interchanges with colleagues I have had while on the faculty of the Harvard Business Schoool, particularly with John Desmond Glover, C. Wickham Skinner, and Curtis H. Jones.

And I want to express my particular thanks to Samuel J. Talucci, who sponsored much of this work, questioned and savored it at every stage, and supported it from start to finish.

Introduction

Implementing Business Strategy

This book is about the role that manufacturing management should play in the planning and implementation of business strategy in industrial corporations. It deals with three management issues. First, it explores means of devising long-term strategic options for the manufacturing function (Part I: Manufacturing Strategy). Second, it suggests methods for strengthening the business strategy by identifying the needs of all support functions (Part II: The Management Row Game). Finally, it discusses techniques for implementing the chosen business strategy by integrating manufacturing and other functional constraints into the planning for a strategic business unit (Part III: Integrating Manufacturing into the Business Strategy).

WHY THIS BOOK HAD TO BE WRITTEN

In my opinion, of all the business functions in the industrial corporation—marketing, sales, engineering, financial accounting, and manufacturing—manufacturing is the least well-integrated into the overall decision-making process.

There are exceptions, of course. In some sectors, such as the automobile industry, it is hardly possible to overlook the importance of manufacturing—even though that industry's performance in the 1970s suggests a failure to keep current with best practice. In other sectors, such as the defense industry, the importance of the manufacturing function is recognized but it is seldom adequately integrated into development plans.

In general, despite all the attention devoted to strategic planning since the 1950s, American industry has failed to incorporate manufacturing considerations adequately into its game plans. The people who have spearheaded strategic planning have seldom had manufacturing experience; their audience, the senior corporate managers, has been increasingly preoccupied with external issues; and the manufacturing managers themselves have not been trained to think in strategic terms, or they have been unwilling or unable to involve themselves actively in the decision-making process until the commitments they will have to live with have been made.

The result of this imbalance is twofold. First, America's manufacturing performance has lagged behind its technical and marketing expertise. Second, strategic business programs have proven difficult to implement.

For a nation that achieved miracles of weapons production during World War II, performed more miracles in mass production for the consumer markets of the 1950s, and pioneered the manufacture of high-technology products in the 1960s, it is a sobering experience to assess the performance of the 1970s. In too many cases, American industry has been unable to match the quality or productivity of foreign competitors, and far too often it took too long to find that out. The disparity in hourly wage rates between American and foreign labor is a valid point, but the aging plant facilities utilized in this country were rarely mentioned as a factor.

By the 1980s, the senior management of most industrial corporations realized they had to deal with the situation, and this decade is witnessing a major restructuring of American industry: plant shutdowns, overseas satellite plants, and automation of existing equipment. What is potentially tragic is that these moves are being taken in some organizations without a restructuring of the decision-making processes that led to the problem in the first place.

WHAT HAS BEEN WRONG IN THE PAST

The top management of American industry has generally regarded manufacturing as a support service rather than as an integral part of the decision-making process. The symptoms of this are:

- cyclical attention to plant modernization
- inadequate manufacturing support for the more mature market segments
- inadequate attention to the difficulties involved in the introduction of new products
- emphasis on short-term results to the detriment of longer-term fundamentals
- preoccupation with financial, legal, and external corporate issues

Cyclical Attention to Plant Modernization

Admittedly, there have been a lot of first-rate new plants laid down in the United States in the past thirty years. My concern has more to do with the upkeep of the plant once it is operating than with the initial capital funding.

When a new business venture is under consideration, the funding and design of plant facilities generally get the full treatment from top management. But once the funds have been allocated and the startup is completed, no one in headquarters seems to want to know very much about the changing needs of the manufacturing function. The result is that the true plant missions evolve and change many times over as marketing people broaden their scope of operations, while the plants themselves become increasingly unresponsive to the new requirements.

The amount of plant funding has always correlated with the ability to pay rather than with the need at hand. Capital investment is high in periods of strong cash flow and low in periods of weak sales. Little is done to modernize plants during the downturns when it could take place in the most cost-effective way; then, when the boom part of the cycle shifts into high gear, there is a scramble for new plants and equipment—just when the machine makers are unable to provide it.

The situation is sometimes even worse. There is a delay factor in the boom cycle that may put the whole plant modernization program into a state of disarray. During the boom's early phase, attention is directed to filling the open plant capacity by laying in materials and hiring workers. When that phase is over, thought is directed toward a better plant. But it takes time to develop sound proposals, so the ideas generally come in during the final phases of the boom, just when senior management is beginning to sniff the advancing odors of a downturn. The result is that many proposals are again delayed or rejected and the plant falls one more cycle behind the product technology and the market forces.

Inadequate Manufacturing Support for the More Mature Market Segments

In the 1960s and 1970s a new conceptual framework began to make itself felt through the planning procedures taken seriously by American industry. Out of some initial studies made at General Electric that were vigorously supported by the Planning Institute for Marketing Strategy (PIMS) and by various research and con- sulting organizations, there came a procedure for defining the different missions of each of the strategic business units in a diversified industrial corporation. The mission statements then became a key factor in the allocation of capital funding.

The concept was called Portfolio Management. It provided a way of looking at the various business units within a corporation in a manner analogous to the way a financial manager looks at his investment portfolio. It correlated market dominance and growth potential with business success and provided a mech- anism called the growth-share matrix to position the business units into different quadrants of an assessment chart (fig. 1).

The underlying concept was basically sound. Evidence indi- cated that a business in a growth market had the potential to absorb the kind of error that would destroy one in a more mature market; that a business with a large share of the market could operate on a more efficient scale in almost every dimension;

1. **Growth-share matrix.**

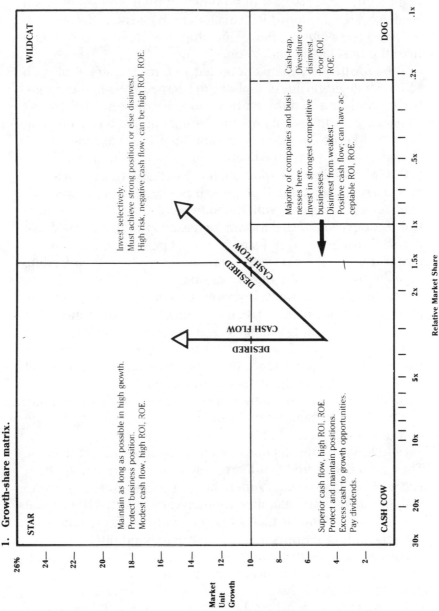

STAR

Maintain as long as possible in high growth.
Protect business position.
Modest cash flow, high ROI, ROE.

WILDCAT

Invest selectively.
Must achieve strong position or else disinvest.
High risk, negative cash flow, can be high ROI, ROE.

CASH COW

Superior cash flow, high ROI, ROE.
Protect and maintain positions.
Excess cash to growth opportunities.
Pay dividends.

DOG

Majority of companies and businesses here.
Invest in strongest competitive businesses.
Disinvest from weakest.
Positive cash flow; can have acceptable ROI, ROE.

Cash-trap.
Divestiture or disinvest.
Poor ROI, ROE.

Market
Unit
Growth

26%
24—
22—
20—
18—
16—
14—
12—
10—
8—
6—
4—
2—

30x 20x 10x 5x 2x 1.5x 1x .5x .2x .1x

Relative Market Share

DESIRED CASH FLOW

CASH FLOW DESIRED

SOURCE: Paine Webber Mitchell Hutchins Inc. Investment Policy, "A Review of Corporate Strategy Analysis," November 19, 1978.

and that the sheer weight of a dominant market position could reinforce strengths for which smaller companies pay dearly, such as brand recognition, price leadership, the ability to obtain and hold the best talent, and so on.

The growth-share matrix soon led to a reassessment of each of the strategic business units in the large diversified corporations—and to a healthy reallocation of support. The "stars" (businesses with both a growth potential and a strong position in the market) were given everything they asked for; the "dogs" (businesses with low growth and a marginal position in the market) were shut down or spun off; the "wildcats" (businesses with low market position but high growth potential) were played like cards in a poker game, with the ante increasing whenever there was a campaign directed toward increased growth; and the "cash cows" (businesses with a strong market position but at the mature end of the growth cycle) were assigned the role of feeding funds to the rest of the organization.

Like all good doctrines, however, this one suffered from misapplication at the hands of its more enthusiastic supporters. For some general business managers, it became a disaster.

It turns out to be injurious to your health to have your business labeled as a cash cow. Once that label is acquired, it is difficult to get capital out of the corporation for plant modernization— or for anything else. In our more swinging corporations the practice has been to bankroll the stars and wildcats and rely on the cash cows to foot the bill.

Unfortunately, the Japanese were playing by different rules. They soon discovered that there were a number of mature markets where the dominant American companies were becoming soft. In one mature field after another, they invested the capital to build state-of-the-art plants and began penetrating American markets with lower-priced and better-quality products.

I am not arguing that entrepreneurial support for growth businesses is bad; I am not even arguing that much of the change in world trade would have taken place anyway, given the recovery of foreign competition after the end of World War II. I am saying something else:

We have been dismayingly slow to understand what was happening to us because we were so bent on playing the risky game of growth marketing that we failed to protect the mature businesses that funded them. We needed managers who would be more vigorous exponents of plant modernization, who could articulate the case for a return on investment in the mature markets that we had taken decades to establish. We needed managers who could visualize the alternative base case: the terrible drift of the cash cow businesses into the class of dogs.

One of the surprises of the last few decades has been the longevity of mature product lines when they are suitably cared for. As with Mark Twain, reports of their demise have often been premature. A bit of my personal experience argues the issue with a rather bizarre example.

> I knew a British carbon company that in 1970 was still making those old-fashioned electric resistors that look like two-inch firecrackers. They had more exciting technology than this, and I suggested that they might get rid of that line in view of the fact that we were now living in a micro-electronic world.
>
> They thanked me cordially for my advice and were rather embarrassed when I dropped in on their resistor plant the following year and found them still at it. They apologized for their dilatory tactics and explained that they had been just about to shut the place down when the sales people discovered a market in Yugoslavia. In succeeding years, they pioneered a remunerative aftermarket in Peru, to keep the ancient municipal trolley cars there in good running condition. Additional markets opened up in Zambia, Paraguay, and, for all I know, Tibet. This time when I visited the place, the plant manager drove me around in a new Rolls-Royce and described their modernization plans with some enthusiasm.

The incident did not make me a believer in obsolete technologies, but it provided a sobering assessment of the longevity in certain markets after somebody declares them to be over the hill.

Inadequate Attention to Difficulties Involved in the Introduction of New Products

Despite America's heavy involvement in new-product technology—and splendid success with much of it—in general we have not realistically faced up to the difficulties of manufacturing effectively for a technology-driven enterprise.

If a product is designed to be at the state of the art, it should also be obvious that it is going to be difficult to manufacture. Apparently, this is not an easy lesson to learn.

The practice has been to give product technology the lead and to let the process technology catch up as best it can. The result is to compound the risk of failure and increase the difficulty of producing a working prototype by adding the challenge of trying to get it produced while the design is still changing and the facilities are inadequate.

This issue is explored in more detail in chapter 6, but it is worth summarizing a few of the problems that arise when an industrial organization tries to spearhead its product technology without taking good care to integrate it with its process technology.

First, inadequate understanding by the product engineers of plant constraints leads to designs that are clumsy, costly to manufacture, or are unworkable without redesign on the plant floor.

Second, incoming parts or subcontracted items cannot be obtained unless they are specifically checked out in advance as available for purchase.

Third, disciplines are ineffective in coping with a continual flood of engineering change orders that are generated after the manufacturing cycle has begun.

Fourth, test and feedback cycles are inadequate to make sure that the product's reliability in field use has been designed and manufactured into it from its very inception.

Last, unrealistic time, cost, and performance commitments are made, all based on the theory that nothing will go wrong during the transition from design engineering to manufacturing.

There is more to the story than this, but the point here is that the problem of new-product introduction is ultimately traceable to an understandable human failing at the general management level: the awareness of the realities necessary to bring off a new technology gets lost in the excitement of the race to be first.

Emphasis on Short-term Results

The great wave of inflation that afflicted the American economy during the 1970s required long-term thinking just at the time when the short-term management pressures were becoming irresistible. Briefly put, our capital-intensive industries had been placed in jeopardy by the overstatement of earnings caused by inflation. The failure of the accounting system to report the true cost of plant in an inflating economy resulted in rates of depreciation that were too low—and taxes that were too high—to encourage plant to be maintained and modernized. Even as far back as the 1950s, it was nine times more costly to lay down new plant in the steel industry than it was to buy going plant through the stock market. The easy way out of this dilemma was to let the heavy industries wear themselves out of business, thus maximizing short-term profits and letting acquisition take the place of modernization.

In the face of the risks of new capital investment and the sure weight of idle capacity, some companies took the seemingly safe way out. Thus the United States Steel Corporation based its new Fairless Works on obsolescent open-hearth technology at a time when the new oxygen inverters were beginning to prove themselves; thus the automobile companies purchased short-term industrial peace by giving the unions whatever they wanted, passing the price along to the consumers just at the time when foreign imports were moving into the market.

This escape route into short-term commitments still lurks behind most long-term strategizing. It confuses and undermines planning beyond what may be necessary to meet the tactical challenges of changing conditions.

In an economic system sensitive to price-earnings ratios (where negative results can lead to predatory take-overs), the quarterly performance review has become a driving force. American industry makes strategies to a fare-thee-well, and then drops the whole game plan whenever the quarterly figures show signs of going below expectations.

What is the reason? Are we terrorized by the stockbrokers? Or are we using these outside financial scorecards as an excuse to put pressure on our own middle management? When General Motors announced that it was passing a dividend in order to invest funds in new plant, it was saying to the financial community, "If you want to sell our stock, go ahead. But you'll regret it when you discover that we've been putting ourselves on firm ground for the future."

Preoccupation with External Issues

With the growth of equity funding after World War II, enormous chunks of capital were placed in the hands of investment-fund managers. Here too the devil took the hindmost, and the fund managers scouted around for interesting growth situations and tax havens. If a company was not onto a high-growth opportunity or it seemed to be plodding along in an uninteresting economic sector, it was penalized with a low price-earnings ratio. It had become easier to acquire assets through stock purchases than to build them with real money, and so the high flyers were able to acquire the low flyers—sometimes with their own cash. The result has been a wave of mergers the likes of which have not been seen since the swinging days of the 1920s.

A good deal of this merger and acquisition activity has been rationalized by that elusive thing called "synergy," a poetic concept that fantasizes remarkable similarities between unrelated activities. If a company makes seats for school buses and it wants to acquire another company that molds taillights for automobiles, it proclaims its entry into the field of transportation. This kind of insight is attractive to fund managers and a good deal of creative literature has been made of it by the people who write corporate annual reports. Several examples of synergy and the actual product lines that caused this magical process to take place are on display in figure 2.

This kind of hype is probably harmless enough, but in too many cases the poets believed in the mythical creatures they had invented. They then called in the manufacturing managers and told them to rationalize the operations. The result was that some were soon trying to reduce manufacturing costs by making bus seats and taillights in the same plant.

As if this were not bad enough, a spree of litigiousness has

2. Corporate synergies.

Proposed Synergy	Actual Product Lines
Electronic and electrical	Electronic cooking units
	Electric motors
	Biomedical research
	Military night vision devices
	Dental equipment
Industrial products and services	Aluminum scrap recycling
	Special-purpose rail cars
	Components for office machines
	Heat-transfer equipment for chemical plants
	Metal stamping presses
Manufacturing	Shock absorbers
	Automobile mufflers
	Construction and farm machinery
	Naval vessels
Energy	Solar roof panels
	Electric toasters
	Voltage regulators
	Small gasoline engines

added nothing to this country's competitiveness in world trade. Product liability cases alone have doubled every five or six years, and the awards for damages have become so fantastic that it is worthwhile to experience some personal damage in order to live a life of ease thereafter. This eagerness to sue has become so nearly commonplace it was refreshing to come across a society without the legal sickness.

I was working in Hong Kong some years ago when a legal issue came up between my client and a major competitor. I supposed that the senior management of both companies would stop whatever else they were doing and go off line for six months to brief the lawyers. But nobody paid much attention to the problem. When I could no longer stand the suspense and asked how they would deal with the issue, the head man (the *tai-koo*) replied: "Oh, that's never a problem around here. There aren't enough lawyers to go around in this place, so we always settle this sort of thing out of court."

Imagine the luxury of having the time to manage the business rather than count the number of summonses in the morning mail.

WHAT THE MANUFACTURING MANAGERS HAVE NOT DONE

If senior corporate management has not been very responsive to the needs of the manufacturing function, it is also true that not many manufacturing managers were willing, or perhaps able, to argue their cases effectively at headquarters. The business colleges and the consulting community have not done much to help them, either.

Schools of business administration have been turning out young men and women smart as whips, eager to make big money in marketing and finance. Where were the graduates willing to go into manufacturing and stay in for the long haul when they were needed?

The consultants meanwhile split into two groups: the grand strategists who worked at the top of the organization, usually innocent of manufacturing experience; and the industrial engineers who dealt with the picky problems that could get budgeted through the production people—time and motion studies, plant layouts, materials requirements systems, line balancing and rebalancing. Not much here to help the manufacturing managers make their case on the big issues of industrial strategy.

The general result has been a particular mind-set back in the plant among the old-guard line managers. It seems to have evolved with the following features:

- A retreat into the hard work of merely getting the product out the door this week with whatever resources they were lucky enough to get
- A fatalism with regard to senior management commitments, in the tradition of the captain who is prepared to go down with his ship rather than demand that the owners provide a more seaworthy vessel. It comes down to an attitude of toughing it out, based on the proud feeling that this manufacturing

gang is able to get product out no matter how ramshackle the factory may be—having long since given up hope of the satisfying feeling that the factory might be the best of its kind anywhere in the world

Getting the Product Out the Door

It is no mean accomplishment in some plants to get the product out the door. However, it is not the only accomplishment that should be taking place.

Managers tend to organize work toward whatever reward system is in place. In manufacturing plants, the traditional performance measurements are units of output per hour, percentage of scrap, indirect charges per direct labor hour, and so on. These tangible measurements are the first things to be reviewed no matter what else has to be done.

They are not the wrong things to measure. The trouble is that in some companies they came to be the only things that were measured. Short-term measures of performance led to short-term orientation by management.

- Indirect charges were shaved down until there was not much staff left to support the longer-term programs.
- Word went out that quarterly targets had to be met, so inventory was shuffled around and the most cost-effective products were scheduled for production whether the market needed them or not.
- Growth factors selected for reports invariably showed improvement over last year.

These systems measured improvement over the previous year, not progress toward a goal. In a period of inflation it was not too difficult to find factors that became larger with each passing year. Unfortunately they may have had little to do with what the business needed to compete effectively in world markets. Undoubtedly some factors required no improvement at all. But there were others for which a 10 percent improvement was total failure because performance had to be bettered by an order of magnitude before the plant could become truly competitive.

Maybe it is manufacturing's macho tradition at work or maybe it is sheer old-fashioned fatalism, but plant managers earn reputations as tough people by accepting extreme commitments at home office meetings and then putting pressure on their people to make do with whatever resources they already have. It is something like the infantry sergeant who tells his men that he did not order artillery support because he wanted to show the old man how tough they are.

Doing things this way can become self-reinforcing. If a plant manager sends in a request for capital funding and sees it rejected, he may update the request and send it in again the next year. But if it is rejected frequently enough, and with enough finality, he is not going to try it again. Headquarters people may wonder why the plant managers make no attempt to modernize their facilities. When asked, the managers explain patiently that this particular organization does not want that sort of thing; they believe that, even if the financial people are now in the mood to approve a few appropriation requests.

Perhaps the most disruptive practice is to chop up a well thought out modernization plan so that it meets some arbitrary constraint. Sound package proposals come back from the home office with some asinine comments such as: "Approved for those line items with a one-year payback" or "Cut the funding in half and resubmit."

Industrial productivity requires a coherent package of related facilities, along with well-supported staff skills and congruent resources, if it is to make any real impact on the manufacturing task. It is simply not possible to put a few pieces of modern equipment inside an obsolete plant and achieve any significant productivity improvement. Manufacturing processes require the infrastructure to support them—and the critical scale to enable them to make an impact on the system. Otherwise, the result will be what has been called "islands of automation" in an environment that cannot use them effectively.

THE MANAGEMENT ROW GAME

I believe that the decade of the 1980s will be seen as a transition period in American industrial organization.

There is a new breed of manufacturing manager moving up the ranks, capable of holding his own in discussions with corporate management—or her own, for that matter, for some of them are well-educated women of the sort never seen in a factory environment a generation ago. There is moreover a responsiveness at corporate headquarters with regard to manufacturing requirements; the onslaught of management articles about industrial productivity and the rise in imported products have generated a new sense of priorities.

Unfortunately, the outcome of this movement is not yet clear. We are, I think, in a race against time.

Simultaneously with all the good things happening to improve America's industrial organizations from the inside, this country's industrial viability is being eroded on the outside by the vagaries of currency exchange rates and the play of international market forces. The outcome would seem rosier if a cavalier attitude did not seem to characterize the wild men and women who acquire corporations to play around with their stock values and the legal people who bring monstrous damage suits against any big corporate name they can find.

In this struggle against time, a conceptual framework is needed to support the good guys—the managers in industry who are trying to help America's industrial base become more competitive on a world scale. Developing such a framework is the objective of this book. Two principles are fundamental.

First, anyone can develop a strategy in business, but only the functional managers can implement it. Unless the marketing, sales, product development, and manufacturing functions have the resources to carry out the attack, it will fail.

Second, the various functions are entirely interdependent. Although they have their own individual constraints and capabilities, and although they speak different business languages, they must act as if they were part of a single organism. The business general manager can expect to be no more effective with a weak sales function, an obsolescent product technology, or a ramshackle plant than an athlete with a paralyzed arm or a broken ankle.

With these principles in mind, business strategy and its implementation can be thought of as a gaming challenge of the most demanding sort. Each of the functions must be allowed the freedom to thrive and pit its strength in competition, but each one is also deployed in active support of the others and the strategic thrust as a whole. This process is a gaming challenge because it is analogous to the kind of skill required of a champion sports team or a chess master.

In a chess game, each piece may be moved in a different way, but all must be played so as to reinforce each other. It is only too easy to deploy the pieces so they get in each other's way or to send them off on some poorly thought out venture without the support necessary to prevent them from being picked off by a well-integrated opposing team.

Just so in an industrial organization. The skill of the general manager lies in the ability to get the pieces working together to reinforce the game plan. It is not enough to have a strategy; it is necessary to play the game tactically so that the pieces can make progress to the objective.

It is possible, then, to think of business strategy and its implementation as being played on a gaming board, with the functional organizations and the business strategies forming a two-dimensional matrix—the Management Row Game.

Under the game's rules, the general manager can select any strategy he chooses, provided it corresponds with the realities of the market in which he is competing. Once he has chosen his strategy, he must see to it that the business functions are lined up so as to support each other. He cannot permit his pieces to become scattered all over the board, with the sales function pursuing one set of priorities and the engineers another while the manufacturing function is kept on half rations, struggling to keep up with both of them.

Figure 3 is a simplified game board with this matrix in place. There are different strategies from which to choose. Each one requires a certain mix of disciplines and activities on the part of the functions if it is going to be supported effectively.

For example, a strategy of market dominance requires high

3. The management row game.

Business Strategy	Marketing Mission	Sales Mission	Engineering Mission	Manufacturing Mission
Market domi-nance	Narrow product line	Price compe-tition	Product stan-dardization	Lowest unit cost
Specialty market niche	Special-ized product line	Premium price	Special prod-uct specifi-cations	Manufacture to specification
Delivery response	Image of dependa-bility	Rapid-delivery service	Design to support changeovers	Produce to changing schedules
Market coverage response	Wide product line	Catalog selling	Broad design support	Produce to small batch runs
Custom product response	Analyze customer needs	Sell on "make to order"	On-going custom design	Introduce a variety of products
Product innovation	Develop growth markets	Sell on prod-uct perfor-mance	Design inno-vation	Manufacture entirely new products
Technical innovation	Assess new perfor-mance require-ments	Sell on technical leadership	Spearhead new technol-ogy	Support new product tech-nology

unit volume production and low unit production costs. This imposes severe constraints on the marketing people who may want to widen the line and encourage new product introduction. It requires a plant that is configured for standardized production and is supported by the kind of mechanization that will be cost-effective for long runs at high volume. On the other hand, a strategy of innovation requires the sales force to obtain the product price levels necessary to support the costs of research. It requires a manufacturing plant flexible enough to respond to continual changes in requirements, to make short runs, and to meet timetables for the introduction of new product specifications.

In each of these strategies, the missions of every one of the functions have to be altered to meet the needs of the business. There are constraints to be met and opportunities to be exploited, but they must be dictated by the business strategy or the plan will not be easily implemented.

This book will examine the Management Row Game more thoroughly by considering how it can be used to incorporate manufacturing into the business strategy. For the manufacturing manager, the objective is to understand the business mission, to interpret the business strategy into manufacturing terms, and to structure the process technology, control systems, and product line support so that the business he serves is at an advantage in competition.

THINGS TO COME

Competitive Manufacturing is divided in three parts.

Part I examines the manufacturing function in its own right rather than as part of a business, to see how the organism works—what it can do and cannot do, and what kinds of capability are at its disposal.

Chapter 1 looks at manufacturing excellence and attempts to summarize some of the factors that make for good or poor manufacturing in different industrial technologies and market conditions. It considers how manufacturing managers can develop a convincing sense of direction for their proposals for plant modernization. Chapter 2 considers how to unscramble the omelette that has resulted from decades of acquisition, diversification, and sheer historical evolution, with the result that manufacturing plants are required to support different and conflicting business objectives. It deals with the task of rationalizing the plant array.

Part II discusses alternative business strategies and determines how the manufacturing function must support them. In the process, the functional interactions that help or hinder the business strategy will also be examined, and what needs to be done to get them working effectively with each other will be considered.

Chapter 3 picks up the concept of the Management Row Game and assesses the needs of the different functions. Chapter 4 examines the marketing-manufacturing interface. The stress points and problems that arise when they are not coordinated effectively are reviewed. Chapter 5 examines the sales-manufacturing interface. The kinds of things that have to be done to make the plants responsive to changing order requirements are determined. Chapter 6 examines the engineering-manufacturing interface, focusing primarily on the transition disciplines necessary to bring new product technology onto the market through the manufacturing facilities.

Finally, Part III builds on the concepts developed, to focus on the problems of implementing business strategy. It presents the results of a workshop process that was developed to bring functional managers together in a structured way so that middle managers can work out their own way of implementing the business strategy.

Chapter 7 reports on the process developed to apply the concept of the Management Row Game to the integration of manufacturing into the business strategy. Chapter 8 assesses the results of the workshop process, reports on the business situations to which it has been applied, and summarizes what it seems able to do. Chapter 9 reviews what has been learned about the integration of manufacturing strategy into the planning and decision-making process. It discusses what needs to be done to make industrial corporations more responsive to the needs of the operating functions that implement business strategy.

Part I

Manufacturing Strategy

1

Developing a Sense of Direction

A senior manager with some manufacturing experience took over a billion dollar industrial corporation after a period when almost no capital funding had been authorized to keep the plants up to date. Response time was slow and the factory floors were choked with inventory. There were something like thirty-five different factories in North America alone. The new president could not spend much time with any one of them, but he asked the plant managers to submit plans for modernization of their operations. This is what he said after he received their proposals.

> I threw them all out! They're going to have to find out what to spend it on before they'll get any capital out of me. Not one of them came in with anything that will lift them out of the rut. All they could think of were new machines of the same kind they have now.

The industrial trade magazines and the brochures of the machine tool companies are full of examples where a particular new piece of equipment saved X percent of something compared with the old process. But that is not what this man wanted.

He wanted a sense of direction: a feeling that the manufacturing managers had a vision of the future, a concept of an ideal plant that they wanted to work toward—something that would pull them beyond the needed improvements that were immediately obvious. He wanted to know what his manufacturing managers were going to do to help the company survive and grow profitably in world competition.

TYPES OF MANUFACTURING EXCELLENCE

It is the business mission that gives the manufacturing function its sense of direction. What can be done to exploit it depends on the excellence of the organization and the resources at its disposal.

But there are two kinds of manufacturing excellence. One has to do with the effectiveness of its support for the business strategy; the other has to do with the efficiency and sophistication of its activities.

These two kinds of excellence can be distinguished through the discussion of three common elements of the manufacturing task: the ability to manufacture to lowest unit cost, meet changing order requirements, and produce new products of advanced technology. The elements will be considered first from the point of view of their relationship to the business strategy, and later from what they indicate with regard to the state of advancement of the manufacturing plant.

Lowest Unit Cost

One possible strategy is to gain market dominance by aggressive price competition. All manufacturing effort is oriented in one way or another to the reduction of cost, so it is always part of the manufacturing mission. This, however, is something different: the deployment of the manufacturing function as a spearhead to enable the business to rely principally on cost efficiencies in order to win an eyeball-to-eyeball price battle against aggressive competitors.

In fact, there are some situations where the business economics itself will dictate a manufacturing effort along these lines—not as aggressive, perhaps, but in the same direction.

These are cases where the process technology leads to a break-even volume that is so high that a business is not even viable unless it gains some substantial share of the market. There is no such thing, for example, as a small styrene monomer plant. Some monomer plants may be smaller than others, but none of them will fit into your backyard.

There are some signposts on the road to manufacturing excellence that can serve as guidelines for the manufacturing managers with this mission.

- Where there is the option to invest in capital facilities so as to reduce the variable costs or to constrain capital investment, the funds should be spent to go after the unit costs.
- Where there are variations in product specifications or elements of the product design that suboptimize production efficiencies, engineering and marketing should know about them as soon as possible so that alternatives can be considered.
- The optimum operating conditions of the plant—the ranges in unit volume and product mix that can be tolerated without significant cost penalties—should be defined so that marketing and sales can determine clearly what those conditions will require of them to sustain low cost production.
- The plant proposed should equal or surpass the best practice in the industry on a worldwide basis.

Flexibility to Change

Not all business strategies rely on price competition to establish their markets, however. A number of strategies depend primarily on the ability of the manufacturing plant to respond to changing market conditions. The business may have a specialty niche in the market, it may offer immediate delivery or a wide range of products, or it may customize the product design for particular customers. In situations like these, while the costs cannot be allowed to drift beyond competitive standards or the prices beyond customer expectations, they may not be the prime consideration. Instead, the manufacturing function must develop a sophisticated system of communications and control that enables it to process and provide a rapid and effective response to a wide variety of orders.

Such business missions tend to generate confusion in the plant because the manufacturing managers are not always sure what kind of excellence is required of them. This may be so for several reasons.

It is difficult for the traditional manufacturing manager to believe in anything beyond controlling unit costs, generating the highest possible volume of production, and getting the product through whatever tests the inspectors have devised. In fact, it is difficult to measure much of anything else. Whatever he is doing, no matter how well he is meeting the objectives of the business, there will always be pressure on him to reduce costs further. He tends to regard the other requirements as an interference to the smooth operation of his plant.

It is also difficult for the marketing and the sales people to define the nature of the flexibility they want or the constraints they want to impose on themselves to make sure that the production costs do not drift beyond what is acceptable. They tend to generate requirements as they can, and let the plant worry about the cost of meeting them.

Interestingly, some newly available technology can provide certain kinds of flexibility without significant cost penalties, in the same way that the technology has begun to improve quality without incurring additional costs. But the procedures have to be designed into the system. The people involved have to know what they are doing and why.

New Product Technology

The product technology becomes the driving force in some other strategies. In the defense sector, for example, elaborate military specifications dominate the process from start to finish. In high-technology industries the product design largely determines what the plant can do. If the product is at the state of the art and unit volume is low, the manufacturing process may be thrown back to the primitive stages of hand assembly. Excellence may consist in merely being able to put the product together more than once.

Process technology exists here too that will help in dealing with some of the new product technologies, but again both must be designed together. The process has to be developed with the specific product technology in mind so that the product can be manufactured in a cost-efficient way.

The relationship between the business mission and the manufacturing mission is represented in figure 1-1. The process, the control system, and the integration of the manufacturing function with the product technology become dominant factors that sometimes overlap in supporting different strategies.

The relationship between the manufacturing function and the business strategy, however, is not the only way to assess manufacturing excellence.

1-1. Manufacturing dimensions of the business missions.

Business Strategy	Process Technology	Control System	Product Technology
Market dominance	Continuous-flow mass production		
Specialty market niche		fully integrated	
Delivery response			
Market coverage response			
Custom product response		control system	producibility
Product innovation			
Technical innovation			of product technology

Effectiveness and Efficiency

What we have been discussing is the effectiveness of manufacturing support to the business mission. There is a different way of looking at manufacturing excellence, however. This way relates to the overall efficiency of the manufacturing operation.

Effectiveness can be assessed according to whether the manufacturing function is a drag on the business mission, dovetails closely with it, or provides a capability through which it can spearhead new market opportunities. Efficiency can be assessed on a more absolute scale less directly related to the needs of the business mission. Moreover, at times manufacturing provides what the business needs but is inefficient; other times manufacturing is operating at peak efficiency but is not in sync with business needs.

Much of this book is concerned with the management task of ensuring effectiveness. This chapter looks at manufacturing more independently of the particular business requirements. The purpose is to help manufacturing managers develop some kind of absolute scale of efficient practice.

TYPES OF EFFICIENCY

The process technology, control system, and product technology are not the kinds of measures of efficiency that would interest an economist.

The industrial economist measures efficiency by assessing the costs of value added. He provides a productivity index by measuring the outputs of a system (total cost less what is paid out for materials and energy and so on) and comparing them with the inputs required in the manufacturing plant (the labor and energy and use of resources and so on).

Instead we will go back to the elements outlined above—the process technology, the control system, and the support for the product technology—to devise a scale designed to help the manufacturing manager determine how advanced he is on each dimension.

Process Technology

The technology of the industry and the nature of the facilities determine what the manufacturing process can do. When the technology permits and the scale of operations supports the capital funding necessary, the process can be very efficient. The epitome of the process at work has some of the characteristics of a modern oil refinery.

In an oil refinery, the product is never touched, shaped, or even seen by the workers—and there are few of them. The raw material pulls itself through the automated process as it works its way to the final product. No work-in-process inventory exists beyond what is necessary to connect the flow from one processing unit to the next. Production scheduling of the plant is conducted through fully integrated process control units so that each operating unit is part of a central system. Sensors and feedback loops in the system anticipate quality problems through the use of statistical control formulas, and automatically regulate the flow or adjust the instruments to forestall them.

A command to the control units changes the rate of flow, or the product mix is picked up by the process units and translated into the hundreds of commands that must be made to adjust the valves, regulate the heat, pressure, rate of flow, and the introduction of catalysts. The units can be programmed to adjust to differences in the type of crude oil used, to determine the best product mix for summer or winter market requirements, and to facilitate the changeover from one to another. The entire system, if programmed and used correctly, can help obtain the best economic results given current prices of materials and products.

There are approximations of these ideal flow processes in the chemicals and paper industries and in certain stages of the steel, aluminum, and glass industries. At the height of Model-T Ford manufacture, operations at the River Rouge plant complex were organized into a single flow from raw materials to finished product, eliminating discontinuities wherever possible. All high-volume, low unit-cost plants tend to set up assembly lines that come as close as possible to a process industry.

But there are limitations in the continuous flow line. It requires substantial capital funding to establish and maintain, which means that a high unit volume of steady demand is necessary to support it. It is so keyed to a particular technology that its ability to adapt to different conditions is limited in that the band width of volume and product mix is narrow. Also, it is difficult to maintain all the stages of the process at the same level of automated flow. In fact, some major attempts to establish the ultimate in production flow plants have become classic examples of inflexibility and the inability to adapt to change in economic or market conditions or in the process technology.

Ford's River Rouge plant, for example, was so fully integrated that it contained its own steel mill, but later, the logistics of materials supply and the cost of labor in Detroit made the mill uneconomic. The Fairless works of United States Steel was entirely based on the latest in 1950s technology, but the new technology of oxygen inverters and steel casting made the huge facility relatively uneconomic.

The purpose of this approach is to guide the manufacturing strategy. By studying industries where the continuous flow process has progressed further, the plant manager can begin to define how far down this road he or she wishes to travel. Not every manufacturing plant should attempt to work toward the model of pure continuous flow. Some plants may be better oriented toward operations where the control system or the product technology determine the nature of the plant.

Control System

The control system dominates the manufacturing capability in a requirements-driven operation where the ability of the plant to respond to changes in the pattern of customer demand is critical. The pattern may be seasonal, such as with sports equipment; it may be cyclic, as with machine tools; or it may be simply difficult to predict, as with toys, games, fashion wear, and "impulse" consumer products. The system may be subject to continual changes in product mix and product design, where the company is providing a wide line of products to a variety of customers.

These are business problems, and the challenge was traditionally met by the business management rather than by the manufacturing plant. The plant simply did its job, and someone else—either the company or the customer or a speculator in between—held the inventory. When production volume was no longer needed, the plant was shut down for a while and its people were thrown out of work.

Three factors have required the plant itself to become more responsive to change. First, it became unacceptable to hire and fire plant workers every time there was a change in the production schedule. Second, there has been a general proliferation of product line offerings as industrial markets have become more advanced technologically and consumer markets more sensitive to style. And third, the increase in "just-in-time" agreements spread the practice of requirements-driven deliveries to many industries that had not needed to operate this way before.

The automobile industry is one of the most difficult industries to which a requirements-driven mission has been applied, due to the complexity of the finished product, the high unit volume, the relatively low unit cost, and the continually changing product design and schedule requirements. Since the stakes are high, capital funding and management effort have been available to advance the technology, and today the auto industry can be taken as a model for other kinds of manufacturing. Much of the discussion of Japanese manufacturing practice comes down to the methods developed in that country to carry out requirements-driven plant missions. This has been a major achievement, considering that they are so far away from their major foreign markets and culturally so removed from them.

The principal means by which manufacturing plants are made responsive to changing customer requirements are the following:

- Very close and integrated logistical planning between the assembly plant and its vendors
- Automation to achieve quality and process control without the need to retrain and supervise workers
- Programmable machine centers designed to provide rapid tool changeovers

- Internal materials management and production scheduling systems designed around minimum process inventories and small batch runs (such as the kanban system, where the ideal lot is a run of one single unit)
- Computerized communications systems to link changing sales patterns with plant scheduling and the whole formidable apparatus of materials disposition throughout the plant

Essentially the process technology associated with these requirements-driven industries comes down to the rather recent effort to further the art of manufacturing flexibility through both hardware and software.

In hardware, programmable machine centers began replacing the large numbers of special-purpose machine tools; robots were developed for difficult-access areas; picking systems were developed for stored components; and mini assembly lines were designed for the insertion of electronic components that were capable of making a different board each time without loss of speed.

In software, the major thrust has been to update the older materials requirements planning systems (MRP-I) for manufacturing resource programming (MRP-II) that incorporates every aspect of the manufacturing planning system from bill of materials to forecasted demand. These systems are driven by product demand schedules. They are intended to be capable of marshaling every needed component to the correct station and therefore of making one product at a time as easily as large runs. They are so complex that expectations of making them fully operational in any given plant run from a few years to a whole decade.

The key point about the requirements-driven process is that the extension of the control system is becoming more of an objective in every manufacturing plant. New technology is available or being developed to provide considerably greater detail control than in the past. The problem for the manufacturing manager is to determine how much of an advance to attempt at any one time.

Product Technology

Product technology in some industries is so advanced that it governs the manufacturing process from beginning to end. The plant may be assigned the task of manufacturing prototypes or complex weapons systems, or of introducing new commercial products out of immature designs. Much of the final design shakedown will take place on the shop floor, and there will be a flood of change orders to cope with. The final product may be so complicated that no test equipment is able to measure its performance fully, and it may have to mate up with equally complex products made at other plants.

This mission has become typical of many American manufacturing plants in the last two decades. These are the conditions under which the Apollo program was managed by NASA and that govern the operations in all the advanced weapons systems contracted by the Department of Defense. They are also the way of life for any high-technology industrial or commercial product, from computers to medical and scientific instrumentation.

The plants in which these products are manufactured are a strange combination of sophisticated and primitive process technology. On the one hand, the products may require such things as: ultraclean areas that look like surgical operating rooms, high bay structures with special jigs and fixtures, very complex computer-testing apparatus and instrumentation, and the safe handling of toxic and exotic materials. On the other hand, they ultimately may be put together by hand by a team of engineers.

The principal mission in such an operation is to cope with extreme specifications and product performance requirements, while at the same time trying to organize the work for the ordinary skill levels of the general work force. Some hardware and software developments meet these needs, and new management disciplines must be learned.

In hardware, the advance has been along the lines of the black box concept, where the objective is to assemble the final product out of a limited number of sealed units, each of which is thoroughly reliable. Instead of leaving the burden of assembly to the last, the objective is to simplify it so that the units can be put together easily and just as easily replaced in the field. The sub-assembly units are therefore designed to make the fullest possible use of components that are themselves sealed and that replace a large number of parts. Typical of this is the very large scale integrated (VLSI) electronic unit, built around the microprocessor, which becomes a building block for the design, manufacture, and servicing of the final product.

In software, the advance has been in areas such as computer-aided design (CAD). These support the design engineer and help optimize and pretest the work. They potentially may constrain the work so that the product is more capable of reliable manufacture and service. The major task is to link the computer-aided design activities with the variety of computer-aided manufacturing (CAD-CAM) systems that may be in the plant.

In management activity, the task is to improve the transition disciplines from design to production. For companies operating at the state of the art, it has become essential to work the problem of the transfer of design from the engineers to the manufacturing team in such a way that the product can be assembled cost-effectively and with complete reliability. This may require new work disciplines at every stage of the procedure, not always an enforceable plan of action.

The manufacturing manager needing to change operations in the direction of a product-driven business will be able to draw on the experience of high-technology industries that both recognized the problems and pioneered in their solutions.

MEASURING MANUFACTURING EXCELLENCE

Every manufacturing plant has to deal with the process technology, the control system, and the product technology. If the manufacturing manager is to set objectives for strategic planning, some kind of scale is needed to help position the plant and assess what else it could become.

Figures 1-2, 1-3, and 1-4 array these factors in a descending scale of nine to one. The process technology scale goes from a manual construction process at the bottom and an oil refinery at the top. The control system has a primitive cost accounting

1-2. Process-driven operations: process technology.

9   Continuous flow process; all units fully integrated; self-compensating process controls; able to flex unit volume and product mix under optimized changeover conditions within a defined range; a "dark" automated factory that needs few people

8   Continuous flow process; separate process units linked by computer to move materials from stage to stage under controlled conditions; process unit is the main element of control

7   Discrete processes linked by automated materials handling so as to approximate a flow process; programming of machine centers is linked to design engineering through an integrated form of CAD-CAM; variable paced conveyor is adjusted by computer to optimize flow

6   Discrete process achieving considerable degree of programmable control; linked through materials handling system that requires manual control; production scheduling through batch lots; paced conveyor

5   Discrete processes using general purpose machines; linked through materials handling system based on kitted parts

4   General-purpose machines grouped into functional departments; batch scheduling by department foremen; material input and output is conveyorized in manual assembly areas

3   Manual assembly; work stations supported by jigs and fixtures; material input and output is supported by separate staff; some general-purpose machine tools; interchangeable parts

2   Manual assembly; based on craft skills; some tooling; "migrant" workers provision their own materials as required

1   Manual assembly; general labor force with undifferentiated skills; each part made to fit on site

service at the bottom and an advanced materials management system at the top. The product technology scale appears to be upside down at first glance. It is arrayed from the most advanced state-of-the-art program at the bottom to a mature and producible product at the top.

The three elements should not be expected to match in any one operation. A plant may be low on the process-technology scale because it makes complex products in batch runs; at the same time, it may be high on the control scale because it has installed a sophisticated materials control system.

It should also be kept in mind that the objective is not always to move to a 9 level. If the nature of the business requires a plant to work at a 6 level, the objective may be to move from a 5 to a 6 or to improve and modernize the 6 level of operations.

The Manufacturing Process

At the bottom of the scale in figure 1-2, level 1 describes a task involving a group of workers engaged in assembly without differentiation of skills. The tools will be multipurpose and obtained from the general pool available within the community. The materials will be drawn as much as possible from local sources, and component parts will be cut on site and made to fit. House-building in pioneer times is an example of this kind of work.

Going up the scale, level 2 involves some differentiation of labor and some use of special tools and materials. Much of the work is organized around craft skills. Apprentices obtain materials for the master builders from a supply depot. Component parts are made by craftsmen who obtain the best raw materials even if it means going beyond the local sources. The construction of sailing ships through the nineteenth century is an example of this kind of work.

At level 3, the mill is power driven and a wide range of machine tools are available. Production is organized as in a general-purpose machine shop. Batch lots are moved through the plant and the machines are set up by experienced machinists. Automobiles were produced this way up to World War I; many machine shops in manufacturing plants were organized on this system until recent times.

At level 4, more attention has been given to each individual work process. Jigs and fixtures have been devised for the work station by industrial engineers. Special tools have been designed for the assembly workers. Materials are moved through the plant by roller conveyors. Before World War II many of the appliance plants were designed this way.

At level 5, the job shop has become a special-purpose man-ufacturing plant with a limited range of products. The emphasis is on low cost and high unit volume. Machines have been de-signed for the type of work required and the product line may have certain standard parts or subassemblies. Production scheduling is planned to optimize the work flow and keep large numbers of workers occupied, though changes in demand lead rather rapidly to hirings and firings. The textile mills of England and New England in the 1820s and 1830s are early examples of this kind of production.

At level 6, the unit volume of standard product is enough to permit continuous runs of the same design. The work stations have been linked by paced conveyors wherever possible so that there are few discontinuities or pauses. The Model-T Ford as it was first made at the Highland Park plant is an example of this kind of manufacture.

At level 7, some machines have become programmable, each one replacing a number of traditional machine tools. Assembly machines have been developed and elaborate test equipment has emerged. Materials stocking is automated and much of the materials flow is undertaken by variable-paced conveyors ad-justed by computer to optimize the work flow between areas. Many present-day plants for appliances and other high-produc-tion products are at this stage.

At level 8, each process unit has become computerized in order to integrate and optimize the activities. Many assembly stations have been taken over by robots. Materials flow between the process units has become highly automated. While the use of robots is limited to specific conditions such as hazardous work, this type of plant is in increasing use.

By level 9, much of the plant has become automated and flex-ibility has been built into the system from beginning to end. As long as the plant stays within its design limits, it can make changes in product mix and unit volume without incurring a heavy cost penalty. It is so heavily automated that it can be re-ferred to as a "dark" factory, without personnel other than those who maintain the equipment and oversee the process controls. There are a few "show" plants like this, but their economic vi-ability has not been generally demonstrated.

The Control System

The control system can be considered the means by which manufacturing management is capable of responding to changes in market demand. The system sometimes advances hand-in-hand with the manufacturing process and sometimes either leads or follows it.

At the most advanced level, designated 9, a total integration of work planning has been achieved and the plant can adapt to changes while optimizing its activities (fig. 1-3).

At the least advanced level, designated 1, historical accounting data keeps management informed; future work is not directed to the optimum channels.

As the scale is ascended, four features in the control system become increasingly apparent.

1. The system changes gradually from an off-line, after-the-fact reporting process to an on-line operational system that can be used to anticipate needs and the actions required to meet them.
2. Separate activities are integrated into a single data base so that information from the plant's different areas is consistent enough to bring all under common control.
3. The control system broadens out beyond the production unit, linking it to materials, product engineering, and forecasts of demand.
4. As more functions are brought into the system, it can be used as a management tool to play "What would happen if?" and to optimize business objectives across the functions.

At level 1, there are no true management controls. Despite voluminous reports from the accounting department, the data is historical. It cannot be used effectively to plan or organize the work or to optimize the plant.

At level 2, the operating and financial accounting data has become useful for planning purposes. However, the data is too much delayed to be used effectively as a guide in controlling operations.

At level 3, the company has begun to develop operating sys-

1-3. Requirements-driven operations: the control system.

9   Fully integrated data-base system; links MIS (financial), CAD, and CAM into a single system; programmable to optimize objectives; re-schedules production to compensate for missing parts or late deliveries; links with selected vendors through requirements contracts, automated ordering and invoicing; links with customers for scheduling of deliveries

8   Same as 9, but without links to vendors or customers; a factory with limited amount of reports on paper

7   Beginning of an integrated control system with considerable manual intervention; a single comprehensive data base, but some difficulty in obtaining disciplined inputs from the users

6   MRP-I system, CAD systems, programmable machine tools, and MIS systems; linkage between them is rationalized to a common data base, but data is incomplete and not remarkably reliable

5   Computer-based systems; but linkage between them difficult and requires considerable "translation" to adjust data bases

4   Mixed computer and manual systems, designed in-house for a wide variety of purposes; much paper generated; overlapping reports and unreconciled data bases

3   Manual financial and operating data; used for management control of production

2   Accounting system and operating data used for planning purposes; limited use of data for operating control

1   Few true operating controls; accounting system provides reports after the fact for historical purposes

tems, but many are still under the control of the foremen, each one of whom may have a different way of operating the shop. Each activity is optimized without knowledge of its interaction with the other activities that make up the manufacturing function.

At level 4, the company has established a formal reporting system that applies to most activities, but the foremen still use their own traditional rules of thumb. Two control systems have emerged—a formal system that integrates much of the data but is largely historical, and a series of hidden informal systems that do in fact direct operations but are not linked to each other.

At level 5, there are computer-based control systems but they have not been rationalized into a single data base. Each system is based on a different set of data. They were developed by "tiger teams" that were eventually disbanded, leaving the systems undocumented and difficult for their successors to build on.

At level 6, control systems have developed that integrate a number of activities but only for certain high-priority programs and certain groups of functions. These systems have proved useful, but the data base is incomplete and not remarkably reliable.

At level 7, the company begins to integrate the computer-aided design system with the materials system, shop scheduling, and product demand forecasts. The control system is linked to the management information system. Considerable work is needed to "clean up" the data base by standardizing product codes and imposing disciplines on the clearance of bills of materials. There is some resistance to these disciplines and the system operates on an uncertain basis.

At level 8, significant management effort has been expended to establish the conceptual framework for a single integrated control system that will eventually encompass all functions. The full extent of this task begins to be realized in terms of the detail and disciplines that will be required and the number of different control systems that will have to be rationalized. Some success is achieved with a materials and scheduling system that pulls together all of the materials and components needs, from a standard bill of materials through purchasing and stockkeeping to the forecasted product demand.

At level 9, the programmable machines that form the computer-aided manufacturing system are linked with computer-aided design and the manufacturing requirements systems. Incoming components and other materials have been standardized and made available. Selected vendors with requirements contracts are integrated into the system so that "just-in-time" deliveries can be made and purchase orders cut by computer links. Similar links are established with the principal customers. The entire system of materials, manufacture, and delivery logistics has been integrated with accounting and planning.

The Significance of New Product Technology

Breathtaking advances have been made in some product technology—to a point where no one knows whether a new product under development will work, let alone how to make it. These advances have interacted with the manufacturing process in strange ways. Sometimes an advance clears away with one stroke a whole series of difficulties that the manufacturing people had been trying to resolve for years; sometimes it throws things back from a high stage of sophistication to some more primitive levels.

Advanced product technology may affect the manufacturing function in one or more of four ways.

1. The new product may be so different from anything made before that it has to be assembled under primitive conditions. For example, the space programs contracted by NASA generally reached a stage where the scientists and engineers themselves had to assemble and test the final product.
2. The new product may incorporate a design that simplifies many difficult components or subassemblies. For example, the seemingly insurmountable difficulties in 1950 of producing reliable electron tubes and electrostatic memories were simply swept away when new computer technology eliminated the need for them.
3. The design of the product itself may have little effect on the manufacturing plant, but the uncertainties of the product's demand pattern may be very disruptive. If the new product is unique or limited to a half-dozen exemplars (each one of which may be somewhat different), there will be no time to develop an experience curve and little budget to organize the work.
4. The difficulties of the manufacturing function may be traceable to a bias in senior management that favors product technology (the life blood of new sales) to the neglect of

process technology (sheer overhead expense). This can lead to cost and quality problems that may well be otherwise avoidable.

It is ironic that product innovation has reduced a number of manufacturing plants to the primitive level of a pioneer house-raising, though without the party of celebration generally connected with it.

Level 1 of figure 1-4 is a program at the state of the art. The design is not yet complete, but it is under such time pressure that it is already in production. Change orders are coming through and more are on the way; the control system has not yet caught up with them, so the manufacturing people are working overtime to make something that will be canceled next week. The design calls for materials or components that turn out to be unavailable, so substitutes are being tested that may require alterations somewhere else in the product assembly. Management has underestimated the difficulties of the new design and the budget has long since been overrun. Vital test or assembly equipment has not yet arrived and may turn out to be inadequate when it does. A critical subcontractor has delivered something to original specifications but it does not meet the current configuration and will have to be redesigned. Prototypes have to be sent out for testing, so key people from the engineering staffs are called in to assemble them by hand. And so on. This situation is not untypical of some advanced weapons systems or high-technology products that are introduced to "wow" the market. What goes on behind the scenes would shock any prospective customer.

At level 2, the program may be difficult but at least an experienced team has been put together to see it through. Its members have been through this before, and while they cannot predict what will go wrong or when, they know they can expect one crisis after another, and they rather enjoy the task of dealing with them.

At level 3, some semblance of management control has taken hold. The program has been carefully thought through at the outset and the critical stages have been identified. Teams have been organized to deal with the most critical problem areas anticipated, and tasks have been scheduled into a critical path

1-4.   **Product-driven operations: product technology.**

9   Sealed subassemblies and components under statistical quality control, assembled automatically by robots or assembly machines; built-in test and diagnostic devices; process under statistical control

8   Simplified product designed for high-speed and reliable automated assembly; modular components capable of plug-in replacement to provide a range of performance options

7   Simplified product, based on producibility considerations; uses standard and reliable components; designed for ease of assembly by regular workers using existing facilities and equipment

6   Complex product, in production; design largely stabilized and value engineered; mature design has been "debugged" and is well documented for producibility; learning curve is flattening out

5   Advanced product, ramping up to production; both engineering and manufacturing functions supported with sophisticated process aids; learning curve beginning to take effect

4   Advanced product, difficult to design and manufacture; but company organized to apply the transition disciplines

3   Advanced product, difficult to design and manufacture; made in a separate part of factory by skilled cadre who will prepare it for a ramp-up to limited production

2   Product at the state of the art, as in 1; but under the management of a business team organized to do this kind of work

1   Product at the state of the art; high degree of uncertainty with regard to performance; subject to continual design changes; performance factors not able to be measured directly; manufacture of prototype unit is conducted by expert team

network for planning purposes. Unfortunately, the plan never catches up with performance once the program is launched because of changes in design and schedule. But at least the planning stage provided the project managers with a conceptual framework to relate to.

At level 4, the program is under some semblance of project

control even though the different elements are poorly integrated. The transition disciplines between engineering and production are being imposed and some effort is made to integrate the product and the process technology.

At level 5, the program is in a more mature phase. It is a Mark II design perhaps, or an advanced version of earlier prototypes. There is some special tooling in the manufacturing plant, a trained cadre of workers, and an established system of suppliers. The program is moving up to limited production and the learning curve is beginning to take effect.

At level 6, design disciplines and manufacturing flexibility have closed in on each other so that change notices can be absorbed in the system. The design has been stabilized and enough unit volume is in prospect to organize the manufacturing effort and train skilled workers on equipment appropriate to the task.

At level 7, the product technology has been simplified to permit the use of special-purpose machines that will help assure reliability and reduce unit costs.

At level 8, the product line has been rationalized and designed for production so that modular units will enable design changes to be made and a wide product line offered without disrupting the plant.

At level 9, the product line has been designed for reliability and ease of production and field maintenance. Sealed components with built-in test capability are extensively used. The process is under statistical quality control, facilitated by precise specifications and accessible test parameters.

SETTING MANUFACTURING OBJECTIVES

The three elements common to all manufacturing plants— the process technology, a control system, and the product technology—make it possible to position each one somewhere on the scale factors just described. With that as a base, it becomes a simpler task to set manufacturing objectives in a productive way. A general-purpose shipyard that builds medium tankers and container ships can serve as an example.

Defining the Manufacturing Task in a Shipyard

The traditional way of organizing shipbuilding is to think of the yard as though it were a permanent construction site: building ways where the structure will be put together, supply depots where the materials are stored and kitted, and general-purpose machine and pipe shops where components are "made to print," that is, manufactured to special design for each ship. It is a level 2 process, organized around cadres of crafts skills and supported with a small army of helpers when the workload increases. The workers are "migrant" in the sense that they walk over to the kitting area to replace materials, go to the storage area to replace a missing component, or sometimes kits are prepared for them by the materials people who bring them to the workplace.

Aboard the ship itself, a situation that can only be described as confusion bordering on calamity may exist: teams of electricians trying to connect wiring in and around other teams of plumbers who are trying to test pipes; welders selected for their small size so they can crawl into confined areas and fuse two plates together, sometimes attempting the infamous "upside" weld that will be subject to reliability problems; foremen and inspectors, carrying blueprints and specifications manuals for reference, wandering around trying to find out what has been done and what has not.

When a shipyard's backlog of orders are not in a delivery crisis and it has a cadre of skilled workers, this appalling process can be organized to achieve some semblance of productivity. The classic example is the Kaiser yard during World War II, which turned out Liberty ships faster than one per week.

Unfortunately, most shipyards do not have a steady enough backlog of work to organize their operations as if they were running a factory rather than a construction site. They are usually either short of work and obliged to make layoffs or they are inundated by work and have to hire large numbers of workers immediately to meet delivery schedules. Some have been known to hire two thousand workers just to be able to keep a thousand because turnover rates are so high during a buildup. This kind of situation can lead to problems:

- lost time while workers look for materials or assemble needed supplies, thus breaking up the work rhythms and interfering with the daily schedules
- interference among teams aboard the ship, delaying work and leading to incomplete welds and poor workmanship
- rework when a unit fails to pass inspection, causing delay and disruption to all the other work around it

Does the scale give any indication of what to work toward? Level 3 mentions jigs and fixtures; level 4 provides for batch setups and kitted parts brought to the workers; level 5 mentions standard parts. But this is a factory here, not a construction site. Even without the luxury of long runs of standard product, it is possible to make a shipyard into a ship factory in three different ways.

1. The ship can be built in sections off-line under factorylike conditions, keeping the craft teams out of each others' way by moving the sections along from station to station.
2. The hull can be designed to make it easier to fit together—with "down" welds made while the section is in one position and "up" welds while it is in another; the work can be sequenced so that the craft teams do not interfere with each other; provisioned kits can be supplied to the work stations; jigs and fixtures to make the assembly more efficient can be provided.
3. The hull can be constructed as if it were a product built out of modules, moving the sections forward by rail or transporter until the whole ship is fitted together and finished.

Ships can be designed for production as if a flexible factory were operating on an outdoor site.

Applying the Method

The purpose of these scales covering manufacturing method, control system, and product-process technology, is to help a manufacturing manager to position himself. They provide a sense of where he is and how far up the scale he might be able to go.

What did the traditional shipyard have by way of a control system? Probably something at level 3, with a planning system overwhelmed by actuality. What should it be working toward? Probably something at level 7, where a sophisticated system marshalls the parts required to kit the work stations and where computer-aided design systems are directly translated onto tapes that will cut steel plates to pattern.

And what is the state of play with regard to the product technology? Probably the yard is at level 5, with boom-and-bust schedules that swing the work force beyond and below optimum levels, and where design change notices interrupt the work flow. If the contract negotiators can quantify these costs and propose alternatives, there may be some hope of establishing a level 6, where the preproduction planning will pay off for the customer.

In other words, shipbuilding is an industry at the 2, 3, 5 level seeking the continuity of work and manufacturing support to organize itself on a 4, 7, 6 basis. With that goal in mind, it should be possible to quantify the business mission and determine what such an operation would look like: what it could achieve by way of operating efficiencies and what it would take to get to that point.

2

Rationalizing the Plant Array

The problem of the plant array must be faced before the task of integrating manufacturing into the business strategy can be dealt with. The modern industrial corporation seldom provides the luxury of one manufacturing plant dedicated to the needs of a single strategic business unit. Usually there are many, each one of which may serve segments of different strategic business units. Every plant is the result of its own past, a jumble of good things and bad, new and old, with history rather than current logic marking the way at every turn.

Consider the typical history of a plant. It is built on a particular site, equipped and organized to carry out a specific mission, and the facilities and services are attuned to a given business strategy. The plant will have the full support and attention of senior management; within reason, it will obtain whatever funding and expertise seem necessary to get it launched successfully.

After the plant is on line, however, traditionally senior management no longer wants to hear anything more about it. The same group that facilitated the plant with the best in machines and equipment will begrudge every penny needed to adapt it to the changing conditions it will have to deal with.

In fact, however, the starting conditions are transient. Once the plant is involved with the evolution of a business, it will be subjected to every change and pressure that the business itself experiences. Several things are likely to happen as the company evolves.

- New products are introduced and product extensions are assigned to the plant.
- The market expands into new end-uses and different types of customers change the service pattern required of the plant.
- The technology gradually changes and the original process equipment cannot meet the new product specifications.
- The product line evolves into two segments: a commodity segment that requires lower unit costs than originally planned into the system, and a specialty segment that requires smaller batch runs and more changeovers than originally planned.
- Another company with a similar but not identical business is acquired, and its products are loaded into the plant.
- The logistics that once favored the plant's location become less favorable as the market demand shifts to new industries or as new raw materials are substituted for the old ones.
- A labor union organizes the plant, or the needs of a changing work force require that management changes its attitudes and develops participative programs that share responsibilities and rewards.
- Vendors and subcontractors once entirely reliable evolve different capabilities and markets and stop servicing the plant.
- Economic conditions periodically force management to break up the support staffs and let experienced workers go.

Soon the plant manager is trying to respond to a dozen different short-term objectives without the resources necessary to deal with them. General management seems to expect a response to all these changes as if the plant were infinitely elastic. The plant manager operates for so long on a "best efforts" basis that he falls into a fire-fighting mode without trying to establish the foundations for manufacturing excellence.

Even if the plant manager were able to plan for the longer term, it would be difficult for him to establish a pattern of excellence without the support of the other functions. No plant can be excellent in all ways. If it is to be the lowest-cost producer, it cannot be expected to provide custom service to different markets; if it is to introduce a continual stream of new products, it cannot be expected to stabilize its work methods.

If military history has taught anything, it is that a winning strategy requires a concentration of force: it is better to penetrate the enemy line on ground of our own choosing than to hold an extensive perimeter with limited resources. This concept has been applied to business strategy and to marketing strategy, and it applies to manufacturing as well.

Plants that have lost their focus are the industry equivalent of the military mistake. The strategic correction is their attempt to define a new sense of mission within the context of the various business strategies they are assigned to support.

THE OMELETTE

What is to be done when the variety of product assignments confuses the mission of a manufacturing plant? To start, the present disposition needs to be understood, along with how it originated and how it fits the future plans of the corporation.

A Policy of Expansion

An outside observer will find it difficult to understand how a particular company came to be the way it is. Most industrial companies are the result of a wild history of change and adaptation—partial modernizations and half-completed facility updates, new product lines thrown in chock-a-block on top of old ones, whole businesses acquired and dumped in (though not assimilated).

There may be half a dozen different plants working at pieces of the jigsaw puzzle, each one trying to service a number of different businesses, though not the same ones at each of the locations.

Sometimes the successful implementation of a business strategy will itself lead to conditions that create a jumble in the array of manufacturing plants, even though every step of the evolution was perfectly logical and necessary.

During the 1970s a major computer corporation expanded aggressively on a worldwide basis. Since the president wanted to establish a responsive organization capable of active technological innovation, he encouraged the establishment of a large number of separate manufacturing plants, each responsible for a segment of the growing product line and each with its own design engineering staff.

By 1980, however, the computer field had changed. It required a closely integrated system with each product closely linked in design and output to the others. Furthermore, the state of the art had evolved so markedly that every capital proposal required significantly increased funding. For example, each new generation of automated test equipment seemed to cost an order of magnitude more than the earlier ones had.

In 1985, the new company president looked over the capital investment proposals and said: "What we need in this company is not more capital funding, but fewer manufacturing plants. I don't care how good they are, we're not going to support every single one of them from now on. I can't tell you how many separate plants we ought to have, but I can tell you that the number is a lot lower than the 50 we have right now."

A Policy of Consolidation

There is a natural tendency to combine operations wherever possible in order to take advantage of the economies of scale. It seems very logical. The plant services can be spread over a larger number of activities, indirect maintenance and support can be allocated on a broader base, and combined operations will enable process units to be properly facilitized for large-scale volume.

However, effective combination requires not only that processes and products be similar but also that they be close enough in process, systems, product technology, and business mission so that they reinforce each other rather than cause interference.

That does not appear to be happening. In many cases the difficulty of managing combined activities is so great and so generally underestimated that it is appropriate to ask at this point whether we should embark on a strategy of separation instead of a strategy of combination. J. C. Heiman of the Eastman Kodak Company presented a paper at a symposium on operations management at the Massachusetts Institute of Technology that illustrates the difficulties.

Heiman starts with a theoretical base of five simple processes

2-1.   Base case: product and process.

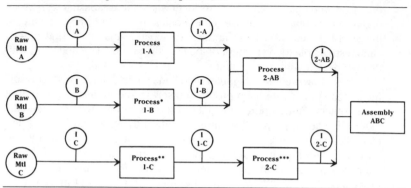

SOURCE: J. C. Heiman, Eastman Kodak Company.

fed by three materials to make one assembly (fig. 2-1). Since four of the five processes are supported by a single prior stage and all five lead to a single assembly operation, the system is easy to manage. There are no setups, because these are dedicated pieces of equipment; there are minimal process inventories and minimal lead times. When any stage in the process experiences a mechanical or quality difficulty, it can be readily identified.

Suppose a product similar to those already being made is to be assembled, using one of the raw materials and three of the same processes as in the first plant (fig. 2-2). The equipment

2-2.   New task: different product using the process.

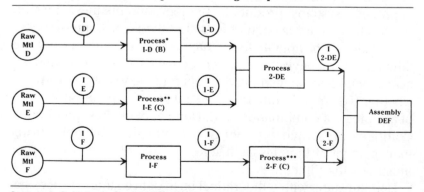

SOURCE: J. C. Heiman, Eastman Kodak Company.

does not need to be duplicated because there is capacity available in the old plant (or a new one can be built to make both products). So two additional process stages and the additional assembly operation are incorporated into the original plant. Rather than having built ten process units, two products come through seven processes.

But consider what has happened (fig. 2-3). Four of the seven processes are fed by two different prior stages or materials and are scheduled by two different assembly operations. Whenever one assembly wants to take over from the other, the equipment must be stopped and reset. There has to be an inventory buffer stage prior to assembly to cover the periods the processes work for the other product. Batches have to be large to minimize setups. Lead times become longer and forecasting and scheduling

2-3.   Effect of consolidating the new task into the existing process.

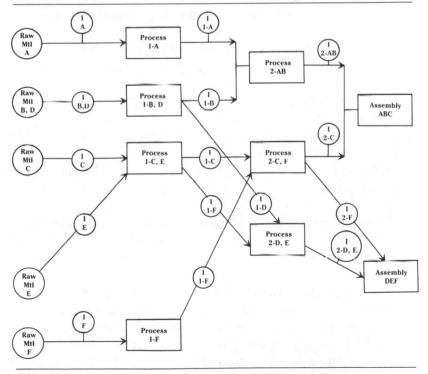

SOURCE: J. C. Heiman, Eastman Kodak Company.

systems will be required. In fact, a sophisticated information system will be needed to manage the production schedule and control in-process inventory. Quality control feedback is delayed and lot traceability may be needed to identify when a particular batch was put through the system.

The end result: the equipment and the work force are utilized more fully—but at the cost of increased complexity in the operating management that may outweigh the savings from consolidation.

CLARIFYING THE PLANT MISSIONS

It has become generally accepted practice in industrial corporations to ask the manufacturing managers to define their plant missions. This is much needed in plants that serve different businesses or where the business needs have been changing rapidly. It is the key to unscrambling the complex situation described above.

But mission statements require a lot of clarification if they are to lead to action. They usually emerge with problems like these:

- Confusion between the plant mission and the mission of those parts of the plant that support one business or another
- Statements that cover so many different factors that they fail to say what the priorities are
- Missions defined in such general terms that there is no way to grasp the focus.

Confusion between the plant mission and the manufacturing mission is probably a sign that the plant has not worked out its organizational structure with the needs of the strategic business units in mind.

In a multipurpose manufacturing plant, the mission may come down to a real-estate and service objective, providing the infrastructure for area manufacturing managers to get on with their work. Each manufacturing manager may have a dedicated facility within the plant that corresponds to the needs of a particular business entity within the corporation. If the businesses have long gone their separate ways and the plant is still functionally organized, it is possible that either the production process is so integrated that it is impossible to separate it into dedicated segments, or the support requirements have not been thought through.

Some missions start with a clear definition of manufacturing priorities and then lose their focus because the planners keep adding qualifications to the statement. "O.K., but we want to produce at lowest unit cost too, don't we?" And "We want to make the highest quality product, don't we?" The result is the loss of any sense of direction the planners may have started with.

But the principal problem with a mission statement is that it may not mean anything. Sometimes a task force of senior manufacturing staff will work up a statement that looks innocuous, yet the staffer will be able to say specifically how he is changing his organization and laying down equipment to carry it out. Sometimes, however, the innocuous-seeming statement is truly innocuous and the manufacturing team does not have any idea how to implement it.

The scale factors developed in chapter 1, and the thinking behind them, can help give the mission statement focus. The method will be applied to a real multiplant-multibusiness situation in an attempt to give the manufacturing operations a configuration that more directly supports the business that it serves.

A CASE STUDY: THE PLANT ARRAY AT NEMCO

The National Electric Meter Company (based on a real company, here given a fictional name) had nine product lines assigned to three plants, in Peoria, Illinois, Ocean City, California, and Gridley, Arkansas. Since their markets were mature, Peoria and Ocean City had experienced difficulty obtaining capital funding for modernization. The Gridley plant was a modern facility close to the limit of its capacity.

The Plants
Peoria

The Peoria plant could be described as a large job shop that had the burden of conflicting missions. It had evolved out of the manufacture of 40,000 medium and 3,500 large electric meters annually for the industrial markets (see fig. 2-4). On top of these traditional product lines, the plant had been assigned two new lines that required precision tooling and skilled assembly. In addition, it also carried out a repair function to recycle old electric meters.

If the equipment had been in good condition, the Peoria plant could have been described as a typical job shop of the 1950s,

2-4.   Interference between the medium and large meters at Peoria.

Medium Meters	Large Meters
800 per week	30 per week
Narrow product line	Wide product line
Continuous requirements	Discontinuous requirements
Broken up into discontinuous batches, to accommodate the work balance	Batches made according to orders, as small as one-offs
Operations could be automated	Operations require hand fitting
But use general purpose machines to accommodate the needs of the large meters	Variety of design and specifications require general-purpose machines
Customers need rapid delivery, which is provided by heavy investment in finished inventory	Customers do not expect rapid delivery
Product line priced to low margins: requires efficient production to be profitable	Product line priced to high margins, suitable to a job shop type of production

making electric meters in batch runs. There was an extensive shop of standard machine tools to make machined components and a large press shop to make stamped parts. The meters themselves were assembled in another portion of the building, where the medium and the large products were made together at certain stages of the process and made separately at others. Space was constricted throughout the plant, and the movement of materials was very difficult since there were many different areas and floor levels. As a result of all these conditions, the in-process inventory was very high and the plant was unable to respond to changes in product mix or unit volume without considerable advance notice.

The control system was also traceable to the practices of the 1950s. The make-or-buy formula kept parts manufacture in-house even though outside sources of supply were available and competitive. The formula was based on standard costs (which were invariably lower than actual costs) and did not include the period burdens.

Quality control was difficult and costly. There was considerable internal scrap and rework. Workers tended to stretch out their tasks because they were inadequately provisioned from the stock room. The factory operated like a series of separate fiefdoms, with each department under the management of a foreman seeking to optimize his own group. The production scheduling system was incapable of ensuring how a new batch would go through the manufacturing process: it was logged into a department, and whenever it was finished it was logged out and sent along to the next stage.

Although the plant had been designed for the manufacture of medium and large meters of traditional design, it recently had been assigned the manufacture of two new instruments. One required careful calibration on special equipment, the other, a new line of turbine meters, required precision machining. One special-purpose machine had been obtained for the turbine line; other components were made by the general-purpose machine shop in the main plant, and rejects were running at almost 30 percent.

Ocean City

The Ocean City plant had become a problem in recent years, with a high labor turnover and a reputation for late delivery. Its sense of mission had been lost in the confusion of product assignments and in the conflict between the character of its work force and the needs of the business.

The plant made standard electrical controls for consumer use. The price level was about ten dollars and the annual volume was approximately half a million units. The product was not made on an assembly line; it was assembled periodically in batches by teams of workers who were then assigned to other duties until another batch run was required.

The plant also made a line of electrical controls for industrial use. In contrast to the consumer products, this constituted a wide product line, custom-engineered to specific use requirements, and assembled by skilled work teams. The average price level was about 80 dollars, and approximately 100,000 units were produced annually for both American and foreign accounts. Sales requirements varied widely from quarter to quarter, and new orders invariably had to be custom designed by the engineering staff.

Ocean City was also assigned the mission of assembling a complex axial meter, designed around a sophisticated investment casting obtained from a single vendor in Chicago. The axial meter cost about 300 dollars and its volume was under 10,000 per year.

Arkansas

In contrast, the Arkansas plant was a fast-moving, cost-effective factory with a lean management organization and a high-volume operation. Its mission was neatly focused.

The Arkansas plant had only one product, a small electric meter selling for about 25 dollars. It produced a million of them annually. This came to approximately 500 per hour on an assembly line that picked up components and subassemblies and conveyed them directly to the final line for assembly, painting, and packaging for shipment. The off-line work stations were carefully engineered with jigs and fixtures and had been timed for a steady but achievable work cycle.

Unit volume was increasing at the rate of 10 percent per year, and the plant was operating at close to its design limits. Further expansion would require both capital funding and an extension of its work force beyond the local township in which it was located.

Assessment of the Product Assignments at NEMCO

The ten product assignments to the three plants at NEMCO have been arrayed in figure 2-5. A simple mission statement is appended to each one.

At Peoria, the problem of conflicting missions was compounded by the characteristics of the different products. The medium meters constituted a narrow product line. With an annual volume of 40,000 they could have been assembled on a modified flow line at a separate plant. Unfortunately, however, the large meters had a very wide product line and required small batch runs, which interrupted the flow line for medium meters and required a different work pace and materials control system to support.

Peoria's second problem was the conflict of technology between the traditional lines of medium and large meters on the one hand and the new instrument and turbine operations on the other. The old meters had bred a type of worker able to fit together a product made to loose tolerances and make it work. The new products required very close conformity to exacting specifications; they could be made only by strictly followed machining and assembly disciplines.

A glance at the ranges in unit value and unit volume will indicate the problem of focusing the plant at Ocean City. Making a half million items costing 10 dollars is orders of magnitude different from making 10,000 items costing 300 dollars each; the cycle time and work pace are different. The differences in technology are even more important: the industrial controls had to

2-5. Product assignments at NEMCO.

Product	Unit Value	Unit Volume	Mission	Support Services
Small meters	25	1,000,000	Mass production to standard design	Industrial engineering for high-volume controls for materials
Medium meters	300	40,000	Flow production to stable design in narrow product range	Controls for production scheduling
Large meters	1,000	3,500	Job shop to stable design in wide product range	Rapid changeover for batch production
Turbine	600	6,000	Precision machining	Modern tooling; skilled machinists
Instruments	800	7,000	Adjustment and calibration	Skilled workers
Meter repair	?	?	Job shop	Product knowledge
Household controls	10	500,000	Mass production to standard design	Industrial and materials controls for mass production
Industrial controls	80	100,000	Job shop to custom design for wide range	Design engineering; job costing; flexibility
Axial meters	300	<10,000	Assembly and test	Skilled workers

be custom designed, while the household controls were standard items and the axial meters required sophisticated machining.

Applying the Scale Factors

The scaling system described earlier can help assess the manufacturing status of each business, define its mission, and indicate what objectives should be assigned to it.

Small Meters

The small meters were manufactured at Arkansas, the only plant with a clear mission. The scaling system assesses it as a 7, 7, 7 manufacturing activity: the manufacturing process was linked by automated materials-handling through a paced conveyor, the control system had at least the beginnings of an integrated control system with a common data base, and the product design had been shaken down for mass production. Improvements were still to be made in all of these areas and the manufacturing staff was working on them, but what had to be done to tighten up the operation was well understood.

Medium Meters

The medium meter line in the Peoria plant was held back from its potential efficiencies by the confused assignments and the terrible state of its support systems. If the unit volume could be increased to the 100,000 level and an adequate manufacturing system developed, the business could be significantly improved.

The manufacture of medium meters probably rates a 4, 3, 6 assessment on the scaling system. The product was moved through the plant in batch lots with kitted parts, using machines that could be adjusted to a wide variety of settings. The control system was primitive and recognized to be so, but the many different product assignments and manufacturing operations made its overhaul a formidable task. The product line was mature and producible even though it had not been rationalized for mass production.

The potential of the business lay in moving the manufacturing operations up to a 6, 7, 7 level. If the product line were rationalized so that it could be assembled on a paced conveyor, the process inventory and both direct and indirect labor costs would be markedly reduced. The question was whether such improvements would help enable the product to reach the 100,000-unit volume level that would justify the investment.

Large Meters

Apparently the manufacturing operation for the large meters was holding back the medium meter line. The unit volume of the large meters was only 3,500; it would always be a job shop operation. It rates 4, 3, 5 on the scale: the machines it used were grouped in functional departments, some of which also supported the medium meter line. This meant that the longer operating times required by the large meters interrupted the flow of the medium meters and delayed them. The control system had been developed for both of them, despite the increased volume that the medium meter line had generated over the years. The product designs, although mature, had never been rationalized for efficient production because of the low volume associated with each one.

Turbines

The turbine line had many of the problems of the large meters but for entirely different reasons. The large meters had been around for decades and required workers who knew the product, while the turbines were new to the company and required some advanced technology. It was a 4, 3, 3 manufacturing operation: the unit volume was still too low to organize a flow line; it depended on the obsolete Peoria control system and was still not debugged for production—the process was not under statistical control and the engineers still had to be called in to help the workmen set up the machines.

Instruments

The instrument line had problems similar to the turbines, although it was largely a manual operation. Instead of the precision machines required by the turbine line, the instruments required skilled workers to test and calibrate the assembled products. It was a 3, 3, 3 manufacturing operation, labor-intensive and slow.

Meter Repair

The repair operation was only slightly more primitive than the instrument line. It was a 3, 2, 3 manufacturing operation: the same labor-intensive activity supported by familiar jigs and fixtures. The control system was different: repair workers logged in their own time and few standards had been applied or made to stick that could be relied upon to set the productivity at a stretch level.

Household Controls

The household-control business had an annual volume of 500,000 units and the product had been pretty well standardized. Yet it had never been profitable in the Ocean City plant. The work force was not geared to mass production and the manufacturing system was nowhere near its potential; in fact, the product was run intermittently on makeshift lines. It was a 6, 4, 7 manufacturing activity: the potential of mass production had not been realized, the control system was barely adequate, and yet the product was ready for high-volume assembly.

It seemed clear that if the household controls could be put into the Arkansas plant, they would have the plant infrastructure they needed to achieve the productivity of a 7, 7, 7 manufacturing system.

Industrial Controls

The industrial-control business had emerged out of the household-control line. They were essentially the same with the industrial products enlarged and customized to meet a wide variety of end-use conditions. Since the product engineering was identical, it had always been assumed that the two businesses should be made in the same plant and on the same machines.

Over time, however, the two lines had drifted farther and farther apart in terms of manufacturing requirements. The household line had been rigorously kept to a standard product, while the industrial line had migrated farther and farther into special applications. The result was that despite its relatively high annual volume of 100,000 units, the manufacturing system was at a 5, 4, 4 level, principally due to the special runs of custom-designed products. If the product line could be rationalized, significant efficiency improvements could be made.

Axial Meters

The axial meter produced in the Ocean City plant was unlike either of the controls products. Only a small amount of precision machining was required to finish an expensive investment casting made outside the plant. The manufacturing operation could probably be rated at 3, 3, 4 level: the in-house part of the value added was not difficult but depended on labor skills to inspect and prepare the purchased part for finishing; the unit volume was too low in this relatively high-volume plant to receive much special attention.

With the scale ratings as summarized in figure 2-6 in mind, it should be possible to array the product assignments according to their manufacturing potential. Rather than making an attempt to improve each manufacturing operation in its present plant context, the ratings will be used to suggest an optimal grouping for each of the manufactured product lines.

Changing the Array

Two objectives are paramount:

1. Put like operations together so that they reinforce rather than interfere with each other.

2-6. Scale ratings at the NEMCO plants.

	Process Technology	Control System	Product Technology
Small meters	7	7	7
Medium meters	4	3	6
Large meters	4	3	5
Turbine	4	3	3
Instruments	3	3	3
Meter repair	3	2	3
Household controls	6	4	7
Industrial controls	5	4	4
Axial meters	3	3	4

The potential:

to improve medium meters to a 6, 7, 7 level

to improve Household Meters to a 7, 7, 7 level

to rationalize the product design of industrial controls

to group like businesses together for improved efficiencies

2. Allow the most advanced activity in each plant to set the standard for the others, so that they will be pulled along more effectively simply by being located within the infrastructure of the more efficient plant.

Aside from the fantasy of a separate plant for each of the product lines, which is not possible, an optimal array for NEMCO might look like the one in figure 2-7. Here the precision ma-

2-7.   Optional plant array at NEMCO.

	Job Shop	Mass Production	Precision Machining	Calibration Assembly
Small meters		X		
Medium meters		X		
Large meters	X			
Turbine			X	
Instruments				X
Meter repair	X			
Household controls		X		
Industrial controls	X			
Axial meters				X

Job shop: Required traditional product knowledge and engineering support; will require complex production scheduling; should be located in a regional city with low labor cost and access to aluminum die casting suppliers.

Mass-production plant: Requires high-speed assembly workers; will need a sophisticated system to ensure adequate flow of incoming materials, and industrial engineers who understand mass production. Should be located in a small town with a reliable labor pool.

Precision machining facility: Requires sophisticated tooling and skilled machinists; should be located near a large industrial city.

Calibration/assembly plant: Requires careful attention to assembly detail; should be located near enough to a city to have access to outside services.

chining facility has been separated from the calibrating assembly plant in order to obtain focused manufacturing missions.

The market volume projected for the turbine business was not able to justify a separate facility. Nevertheless the plant array has been rationalized, and in figure 2-8, the scale ratings for a more practical array are set forth.

One plant has become the job shop for the entire business. It handles the large meters, the industrial controls, and the repairs for the traditional product lines. Such a plant would organize itself around batch production, with an engineering staff to support preproduction planning and with the support staff necessary to establish the control systems for scheduling a wide range of products.

A second plant is the mass-production facility that makes the high-volume products, both small meters and household controls. If medium meters are put into this plant as well, the fast cycle time would reduce manufacturing costs and support a growth strategy for the line.

A third plant combines the precision machining of turbine meters and axial meters with the skilled assembly of the instru-

2.8 Scale ratings for the optimal plant array.

	Process	**Control**	**Product**
The job shop			
Medium meters	5	4	6
Large meters	4	4	5
Meter repair	3	3	3
Mass-production facility			
Small meters	7	7	7
Household controls	7	6	7
High-skill facility			
Turbine	4	3	3
Instruments	4	3	3
Industrial controls*	4	4	4
Axial meters	4	3	3

*Note that the industrial controls could go either into the high-skill facility or into the mass production facility, depending on the degree to which the marketing people felt they could standardize the product line. The assignment here indicates that the marketing people refused to limit the range of the industrial controls products.

ment line. This would also be the best location for a design group assigned to the development of new products.

Figure 2-9 presents the final rationalization plan.

Peoria keeps to its traditional meters, both medium and large, and to the repair function. The Arkansas plant is given the household controls as well as the small meters, and is expanded and modernized. The Ocean City plant is shut down. A new plant

2-9. Rationalized plant array at NEMCO.

	Peoria	Arkansas	New Plant
Small meters		XX	
Medium meters	XX		
Large meters	X		
Turbine			XX
Instruments			X
Meter repair	X		
Household controls		X	
Industrial controls			X
Axial meters			X

Peoria mission: A job shop with some dedicated equipment. Objective is to reduce manufacturing costs of its principal product line, the medium meters. General-purpose equipment to back up the other two product lines.

Arkansas mission: A mass-production plant designed for maximum cost-effectiveness of its principal product line, the small meters. Objective is also to "bootstrap" the household controls to improved cost performance through mass-production methods.

New plant: Space to be leased near a major industrial city. Objective is to obtain a cadre of skilled machinists to support the turbine line. Skilled workers should also be available to calibrate instruments. Design engineering principally dedicated to the turbine line, but able to support the industrial controls business as well.

is required, to make turbines, industrial controls, and instruments in a region close to the vendor of investment castings and to a labor pool of workers familiar with modern machining.

In order to make these new product assignments effective, a concept of "principal mission" was defined for each plant to give it direction. Peoria was held to a job shop but with the principal mission of making the medium meter line cost-effective with whatever dedicated equipment it required; Arkansas was strictly mass production; and the new plant supported the higher technology products.

UNSCRAMBLING THE OMELETTE

Well, nobody can unscramble an omelette.

Combinations and consolidations have gone too far to be un-done. The cost of central services, from plant clean rooms to materials-control systems, has risen too high to allow for much duplication of the support functions all the way down the line. This is true even without consideration of the logistics that may well anchor a product line to a particular place for reasons of proximity to critical materials and markets. However, as the NEMCO case suggests, it is possible to apply a methodology that goes beyond general statements of plant missions, so as to help define what the present manufacturing is and what its potential for growth might be.

Sometimes there are surprises in what should go together and what should be kept apart. In the NEMCO situation, for example, the industrial controls resembled large versions of the household controls: the product design was similar. The manufacturing needs of the two product lines were entirely different, however: the design of the household controls had been frozen to allow for the efficiencies of mass production, while the industrial controls were custom designs that had to be engineered for small-batch production. At the same time, the instruments and the axial meters had no common design elements whatever, but they both required careful work to calibrate, measure, and assemble: the work skills could reinforce each other.

With judicious placement of product it is possible to give a boost to some manufacturing activities by placing them with a more advanced plant higher up the scale. Bearing such a possibility in mind, these analyses should always look to the future. The trends and competitive pressures need to be known, as well as where the manufacturing function will have to advance if it is to support the business aggressively in competition.

In the NEMCO case, an issue emerged with regard to the future of the industrial controls line. If the market plans were to rationalize the line so as to reduce costs, it would have been best assigned to a plant capable of flow production. If the plans were to maintain or emphasize the custom engineering function, it would lead to a job shop mission.

The assignment of household controls to Arkansas gave them a kind of free ride on the mass-production expertise of the plant. Almost everything done to improve the mass production of small meters also improved the production of the small controls—the work force understood the requirements, the control system provided the materials, and the business helped management level the work load throughout the plant. The 7 level of the household meters was able to pull the controls line out of the general service plant it had been placed in and give it a new level of manufacturing efficiency.

If it had been possible to rationalize the large-meter product line of the Peoria plant and increase unit volume by about 30 percent, the meters could have been assigned directly to the modified flow line set up for the medium meters.

SUMMARY

The NEMCO case illustrates both the difficulties and the compromises required for their solution. What essentially has been structured with all the multibusiness plant consolidations is an inherent conflict among the plant manager, the general business managers, and the manufacturing managers that serve each of the business units.

The plant manager is trying to operate the plant as consistently as possible. He would prefer a single business with a single mission; in lieu of that he tries to be as responsive as he can to the needs of the different business managers. But his operation is not infinitely elastic, and he has a responsibility to maintain some kind of coherent mission for the plant as a whole.

Each business manager wants as much response from the manufacturing manager as possible. He would prefer a plant dedicated to his own business; in lieu of that he wants an understanding of his priorities and as much service as the plant can provide.

The manufacturing manager is the man in the middle. He is trying to make the operations under his control responsive to the different business managers he works with, but he depends on the plant manager for support services. He may have three or four different product lines or businesses under his jurisdiction, so he has a kind of multibusiness plant of his own to run within the framework of the main plant.

Part II

The Management Row Game

3

Playing the Management Row Game

The Management Row Game is not really a game—the economy is at stake in this task of implementing business strategy. But it has its rules, like chess.

In chess, the strength of the pieces lies in their relationships to each other, not in their individual positions on the board. Out of eagerness to win, a player can send out a queen unsupported and lose her; but with care even the pawns can become formidable. A good player sees the connections between the pieces on the board—horizontal, vertical, and diagonal lines of force reaching out from the rooks and bishops, and the asymmetrical landing points of the knights.

In business competition, the success of a strategy depends on how well the functional missions are carried out. The pieces the general manager deploys are the functional resources—marketing, sales, engineering, and manufacturing—with financial accounting keeping count of the moves. It is a row game, because the functions reinforce each other when they are properly lined up to carry out a particular business strategy.

The danger in American industrial management is a wild eagerness to win fast, a style of play that tends to propel these functions into a competitive world without considering how they will be supported once they get there.

STRATEGIES

Definitions come first.

A *business function* is an organizational unit comprised of a professional skill such as marketing, sales, engineering, or manufacturing. Each function requires a certain organization, the resources to carry out its mission, and the communications necessary to link its work with that of the others.

A *mission* is a comprehensive task assigned to the business function that enables it to focus its resources and organization on specific long-range objectives, creating the culture and developing the skills and emphasis necessary to enable the business function to carry out its part of the strategy.

A *strategy* is a particular way of approaching a business objective. Every separable business must have some kind of strategy; each of the functions has a strategy as well, whether it is clearly defined or not. A strategy may be aggressive or cautious, it may rely heavily or lightly on technology; it may require much or little funding; it may have a high or low level of risk. Whatever the business strategy, it has to be translated into action through the business functions.

The concern here will not be so much with what strategy a business unit ought to develop but rather with what functional support will be needed to implement one. Plant missions and manufacturing strategy are focused on, to ensure that these considerations are thoroughly integrated into the business thinking.

The Management Row Game is laid out in figure 3-1. Some of the possible business strategies are arrayed in rows, while the four principal functional missions needed to carry them out are listed in columns. The game will be approached from the point of view of the general manager—either the owner of a small to medium industrial business or the profit-center manager of a coherent business segment within a larger corporation. It is assumed that a business strategy has already been formulated and the manager now has the responsibility of making sure that the organization will carry it out. The strategies identified in figure 3-1 will be considered as well as what the functional units will have to do to support them.

3-1. The management row game.

Business Strategy	Marketing Mission	Sales Mission	Engineering Mission	Manufacturing Mission
Market dominance	Narrow product line	Price competition	Product standardization	Lowest unit cost
Specialty market niche	Specialized product line	Premium price	Special product specifications	Manufacture to specification
Delivery response	Image of dependability	Rapid delivery service	Design to support changeovers	Produce to changing schedules
Market coverage response	Wide product line	Catalog selling	Broad design support	Produce to small batch runs
Custom product response	Analyze customer needs	Sell on "make to order"	On-going custom design	Introduce a variety of products
Product innovation	Develop growth markets	Sell on product performance	Design innovation	Manufacture entirely new products
Technical innovation	Assess new performance requirements	Sell on technical leadership	Spearhead new technology	Support new product technology

Market Dominance

One of the more interesting findings in recent market studies is that businesses with dominant positions in their markets generally enjoy higher profitability than businesses with marginal market positions. This can be attributed to a number of factors: the price and volume stability of the market leader; the production efficiencies of manufacturing on a large scale; the power of an established name in the marketplace; the ease of selling from a position of strength; and the sheer momentum of an established position, which makes it possible to plan capital investment for a longer period with confidence.

This concept of market dominance attracted a good deal of attention in the 1960s, together with some convincing examples that ranged from the history of the Model-T to the positions achieved by IBM and Xerox in the world markets. The results were mixed, however. A number of companies that did not enjoy market dominance undertook aggressive campaigns of price cutting to achieve it. It turns out to be a high-risk approach to business competition: there is a considerable difference between strengthening an established position and breaking into a fortress.

In order to parlay a small market position into a dominant share, a business has to gamble heavily on the future. It has to think through its operational planning in such detail and back it with such heavy capital funding that it can bring about the market coup by sheer weight of effort.

The most successful practitioners of this art have been the Japanese. Conditions were favorable for this kind of strategy in the 1960s and 1970s because the Japanese domestic markets were protected, orderly low-cost financing was available, a cooperative work force was present, and considerable disciplines were enforceable throughout the system of supply to make sure that the product displayed superior performance.

The strategy involves the steps that follow.

1. Undertake thorough market research, product development, and quality control to ensure a reliable, desirable, and cost-effective product.
2. Provide whatever capital investment is necessary to establish the best large-scale manufacturing resources, vendor network, and materials supply, so as to ensure low unit-cost production.

3. Build up the domestic market position to establish the home base necessary to support penetration in the export markets.
4. Price the product aggressively, starting with whatever it takes to penetrate the market and continuing with pricing levels guided by a projection of future experience curves.

This is a dangerous process as a number of businesses have discovered, and it requires careful implementation. The company must be prepared both to lay down whatever capital funding may be necessary to stay in the game and to price aggressively enough to penetrate the market without either setting off a price war or going below the cost performance levels that the manufacturing people can support. It is all done with mirrors: if the marketing people price the product low enough they will generate market volume; and if the volume comes in on schedule, the plant will have to be able to reduce its costs enough to make money.

The manufacturing function has to plan all its operations for a level of activity far beyond what it currently has to support this strategy. The logistical system, from incoming materials and components to outgoing product, must be organized for a high-volume level; the plant has to be facilitize with high-volume production equipment; workers must be trained to become a cadre around which additional workers can be brought in and trained in turn; and a system needs to be established that will ensure quality of the product at high rates of assembly.

The manufacturing function also has to know what it needs from the other functions and make sure that it gets it. From design engineering, it requires the closest possible coordination during the introductory phase so that the new product will be producible as far as possible through automated methods of assembly. From marketing, it requires the disciplines to keep to standard products without widening the line beyond the optimum limits of the plant. From sales, it requires the drive to obtain the necessary unit volume of orders without creating a disorderly market.

Internal pressures on the organization are tremendous, and they must be anticipated and planned for. The early years will be financially disastrous because the plant investment must be front-ended and inventories built up before sales are made: the company will be operating ahead of itself throughout its growth cycle. Pricing can get out of hand at any moment and bring the whole show down. Above all, the manufacturing function has to deliver on its promises all the way through from cost containment to product reliability and delivery schedules.

In order to carry out such a program, everyone connected with the business must have a single-minded purpose. They cannot let themselves be deflected into broadening the line or updating the product before they absolutely must. In short, they cannot follow other missions than the one laid down for them.

I make the point about single-mindedness, because almost every manufacturing mission I have ever seen contains the phrase: "And we will produce at the lowest unit cost." But if the manufacturing function is serious about a mission to achieve lowest cost, it cannot carry out a lot of other missions at the same time. If most mission statements were more truthful they would read: "We will produce at the lowest cost we can, while providing the responsiveness to other objectives you want us to carry out."

It is simply not possible to implement a business strategy that can be translated to mean: "We're not going to modernize our manufacturing facilities and we're not going to constrain the other functions from imposing conflicting objectives on the plant. But we want to have the lowest unit cost anyway, because we may want to cut the price whenever we think we need to in order to meet our growth targets."

The Special-market Niche

A business does not have to dominate the market to be profitable. It can get such a convincing grip on a special segment of the market that it can command a price premium. The strategy requires a different set of functional missions than the strategy of market dominance.

To begin with, the specialty markets have a smaller population base and they tend to experience wider swings in demand than the major markets. Some of the niches lead to luxury or postponable purchases, which are sensitive to economic shifts or style changes. The markets require careful monitoring to keep in tune with demand. Above all, specialty niches require careful tuning by the marketing people and the design engineers to meet the changing needs of the market.

From a manufacturing point of view, this strategy places less pressure on unit costs than one of market dominance: if there is any market out there at all, it should be able to support a wider margin than the commodity products. But it will have to be catered to.

The defense market is such a specialty segment for companies that make components with a major industrial requirement, such as electrical connectors. For the military segment, the company will be obliged to manufacture products that meet a battery of special tests and requirements, from the way they are labeled to the way they are packaged and shipped. The high-priced end of the automobile industry is another such segment, where the customer expects to get something unique in design or performance for his money. Specialty papers and chemicals and alloy steels are also special markets, in which the products justify their higher price by meeting more exacting tolerances or providing more value added than required by the commodity portion of the market.

For the manufacturing function, specialty markets require very careful attention to detail over batch runs that may not be very large. Without the volume to support high unit production, the process has to be designed to provide the quality necessary to support the product line. This means close work with suppliers to define the specifications of incoming components and materials; it may require certain pieces of automated equipment, not so much to reduce labor cost as to ensure quality controls; it will certainly mean the training of skilled workers and their supervision.

A good deal of interaction with the product engineers will be required of the manufacturing function to make sure that both functions have identified the most critical aspects of the design on the shop floor. Procedures will have to be developed to make sure that close tolerances can be met, and systems put in place to enforce disciplines.

From a business point of view, the strategy requires continued attention to make sure that the products justify their higher prices. This may mean continual upgrading of specifications and continual pressure on both design engineering and manufacturing to advance the state of the art.

Responsiveness

Another business strategy can be characterized by the motto: "Give Them What They Want." Some people describe it as no strategy at all, in the sense that there may be little pro-active drive toward particular objectives: the company's annual report may rationalize a good deal of opportunism as broad market coverage. On the other hand, the strategy may result from a conscious effort to be responsive to market needs on the broadest possible front.

Whatever is behind it, opportunism or design, a strategy of responsiveness is in fact a widespread approach to business growth, one that has great consequences for the manufacturing function.

> A defense contractor has grown five times over in as many years. Management wants to improve productivity. A close look at the contracts they have taken on discloses a hundred programs of every size, type, and technology imaginable—small tanks, electronics equipment, large-scale aircraft support programs, mass-production ammunition contracts, and special vehicles. Prime contracts, subcontracts, and parts manufacture—whatever it is, they will do it because they are afraid to lose anything. Management is organized to take potshots at whatever flies across the horizon, and then they tell their manufacturing people to make it.

How is productivity going to be increased? By getting them to stop bidding on everything within sight.

Somewhere buried inside these exuberant growth commitments is a valid strategy of responsiveness that can be defined, constrained, and implemented. Referring again to figure 3-1, responsiveness can be seen from three points of view—delivery, market coverage, and custom design.

Delivery

Delivery responsiveness is a strategy that says to customers: "You won't need to hold inventory. We'll meet your needs no matter how they may change." This loose statement must be defined carefully before it can be made to work. It probably should be translated to: "You can hold a minimum inventory, and provided you keep us informed of your changing needs, we will do our best to meet them with less delay than our competitors."

In order to make this strategy effective, the product line must be limited to those products that can be process engineered for a flexible response. Then the factors that make up the lead time between the company and its customers have to be thoroughly understood and agreed on by both. Suppliers of component parts and raw materials must be mobilized. Process equipment and labor hours have to be thoroughly prepared. A master control system should be in place to control the parts flow and make the relevant production resources available when needed.

Planning like this will require detailed communications from the manufacturing function to the other functional managers so they will know what can and cannot be done. It also requires extensive communication with the principal customers (and suppliers) to coordinate requirements if large finished inventories are to be avoided. The strategy will impose some disciplines on the salespeople so that they check back with the plant and determine the current work status before they make commitments to new customers.

Market Coverage

A strategy of broad market coverage can be supported if the product line is defined and needed resources made available. Broad, however, is not infinite.

Product proliferation causes a slow deterioration of manufacturing effectiveness that is difficult to recognize, measure, or correct. Because it reduces the batch size of a typical manufacturing run, it has the effects that follow.

1. It increases the number of changeovers or setups. This may be a minor problem if the equipment is programmable and designed for flexibility; it may be a major problem if the equipment has to be cleaned out from the old run or set up by hand.
2. It reduces learning curve efficiencies, because the work team has to start up again from a dead stop.
3. It increases the possibility of quality problems, because some of the problem products will be run at infrequent intervals.
4. It causes much difficulty in production scheduling and inventory management, increasing the inventory-to-sales ratio markedly in some cases.

Most of these costs are not entrapped by the regular cost-accounting system in the plant, and many of them are difficult to measure even with special studies. Often it is only after the drift has become established and performance has deteriorated that the issue is brought into the open.

Custom Design

Some companies are organized to offer custom-designed products or components to their customers. This works well where the commitments are within the range that the manufacturing plant can respond to. It can get out of hand very easily, however, when an enthusiastic salesman (who is sometimes the owner or president) makes commitments without clearing them with the manufacturing manager. In one form or another, the following situations can arise.

- The product can be produced, but not within the time scheduled for delivery unless other products are rescheduled and delayed.
- The product requires some special part or treatment that needs to be subcontracted outside the regular sources of supply, causing both schedule and quality problems.
- The product can be made, but the unfamiliar, difficult specifications and close tolerances require special attention; the

first runs have to be scrapped or reworked before the plant can get the job done correctly.
- The product looks readily made by the equipment within the plant, but careful examination of the specifications and design indicates that it will require special process equipment that the plant does not have—and no one knows who does have it or if it can be found.

The most critical skill of a custom house—whether making metal cabinets for industrial products or major weapons systems for the Department of Defense—lies in its ability to bring together the sales commitments and the manufacturing resources. When this link breaks down, there can only be trouble.

Technological Innovation

The swing toward high technology has been so much publicized that it is easy to forget how difficult it is to manage successfully. Innovation, which highlights the design engineering function, depends on successful financing, marketing, and manufacturing to sustain it. The roadside is littered with the corpses of businesses that built better mousetraps without providing the capital funding and viable product lines needed to sustain a cost-effective manufacturing function once the initial mouse had made its way into the trap.

Since such companies are driven by entrepreneurs of the technological persuasion, the chances that they will develop a disciplined manufacturing organization are not high. If a manufacturing manager is put in charge of any of these companies, it will be brought to a dead halt; but if he is run over roughshod, the new products will be a shambles before they get out the door.

The author of this book researched the field in the 1960s in preparation for another booked called *The Management Problems of Diversification.* The study was focused on the problems of a dozen or more high-technology growth companies that had been acquired by established manufacturing companies, most in the business of manufacturing subassemblies or components in the automotive industry. The logic was clear: on the one hand, there was the high-technology company with growth potential but inadequate finances and production know-how; on the other, there was the established manufacturing company with capital and production skills but no growth prospects. Get them together, and the best will be drawn out of both worlds. In fact, things generally did not happen that way.

When the high-technology company was placed under the jurisdiction of the manufacturing corporation, pressure built up to freeze designs and get into production. But in most cases the designs were not ready to be frozen and were in no shape to be produced. Also, in several instances the market was not ready for the product in the numbers that the plant wanted to turn out. Everybody was still feeling his way with the new technology. The disparity was too great between the high complexity and low unit volume that the technologists were used to and the standardization and high unit volume that the manufacturers were used to.

PLAYING THE ROW GAME TO LOSE

It should be clear by now that there are a number of ways to lose the row game. Of course, even an expert style is no guarantee of a win in a market economy where there are always surprises. But enough evidence has accumulated to identify some of the ways of making sure that the game is lost early and with some panache.

A Loose Style of Play: Migration of the Functions

Each of the functions defined in figure 3-1 has its own organization, culture, and technology. Functions left to their own devices tend to optimize their own potential. Each one has a model, consciously defined or not, of how it should ideally operate. When not tightly constrained by the needs of the business, each will migrate in the direction of its ideal.

Unfortunately, each migration tends to move in a different direction. The result is a scattering of the functional pieces all over the row game board.

Left to its own devices, the marketing team will migrate down the board in the direction of innovation. Up at the top of the board, the concept of market dominance and market niche may appeal to a group trying to establish itself. But these areas of play require a good deal of discipline from the marketing team. Once a product or line has been established, they cannot do much with it. The only way for them to identify new opportunities and define new product needs is to move down the game board into responsiveness and innovation.

The sales team is bimodal. On the one hand, it likes to operate in the responsiveness sector; here the salespeople can offer whatever products or services the customer may need. Almost every call results in an order, provided only that the customer has some need still unattended. On the other hand, the sales team likes to reach up into the sector of market dominance wherever necessary and offer a price discount to confirm the sale.

The engineering team clearly wants to work at the bottom of the board in the innovation sector, and only in that sector. No one wants to work in the restrictive disciplines of market dominance, where standardized product is required, or even in a niche environment, where updating and extending the existing product line is all that is required. A good team has to be allowed to play about in the innovation region or they will soon be looking for more stimulating jobs elsewhere.

Meanwhile, the manufacturing team is trying its best to migrate up to the top of the board to the market dominance sector. Every manufacturing manager has an image in mind of a factory that operates as if it were one single machine. In come the raw materials and component parts to a plant sector that identifies, stores, and picks them; off they go to be processed in one linked stage after another with no stops and starts; and out the door they go, packaged and labeled by another part of the machine and carried off to waiting customers by a line of attendant vehicles. The role of the manufacturing manager, in this ideal image, is to tune up the machine, keep it oiled and greased and moving as if it were an automobile engine getting ready for the Indianapolis 500.

These migratory tendencies are identified in figure 3-2. They

3-2.   Migration of functions.

Business Strategy	Marketing Mission	Sales Mission	Engineering Mission	Manufacturing Mission
Market dominance		price concessions		uninterrupted production flow
Specialty market niche				
Delivery response		rapid response		
Market coverage response		wide coverage		
Custom product response		custom design		
Product innovation	market leadership			
Technical innovation			exploration of new technology	

indicate what will happen to an organization if the general manager indulges in a loose style of play.

This kind of looseness has unfortunately become mixed up in the minds of senior managers with the concept of delegation of authority. Delegation was instituted for the purpose of getting senior managers out of the details of the functional activities, so that the general managers and the functional managers could get on with their work. But delegation is not the same thing as abdication. Delegation can only be properly applied when both sides, the delegator and the delegatee, know what is being delegated and what is not. And one thing that is not delegatable is the game plan. It may be the result of a good deal of consultation among all the managers, but it cannot be left to the migratory preferences of each functional manager.

Changing the Signals: The Time Factors in Response

A new strategy or a tightening of performance under the old strategy has to take place whenever economic conditions force a change in strategy or new technology sweeps away the old standards of acceptable performance. However, each function requires a different amount of effort and time to respond to new conditions. For this reason, a change under stress may break them apart. This disarray is represented in figures 3-3, 3-4, and 3-5.

Take as an example a business launched with a strategy of market dominance. Over the years it has lost its focus and spread out in a de facto strategy of market coverage: the marketing function has encouraged product proliferation in an attempt to expand markets; the sales function has something to sell from that looks very much like a catalogue, but is giving discounts to hold its customers; the engineering department has been supporting the general drift by making extensions to the product line; and although the plant manager has been told to maintain

3-3. Changes in strategy: the natural drift toward product proliferation.

Business Mission	Market- ing	Sales	Engineering	Manufactur- ing
Market dominance ↓	Narrow product line ↓	Price competition ↑	Product standardization ↓	Lowest unit cost ↓
Market coverage	Wide product line	Catalog selling	Customized product design	Batch runs and Frequent changeovers

3-4.   Changes in strategy: attempt to rationalize the product line.

Business Mission	Marketing	Sales	Engineering	Manufacturing
Market dominance	Narrow product line	Price competition	Product standardization	Lowest unit cost
Market coverage	Wide product line	Catalog selling	Customized product design	Batch runs and Frequent changeovers
Market niche	Rationalized product line	Stabilized pricing at lower levels	Modular design	Lower unit cost objectives

low unit costs, the small batch runs and frequent changeovers have actually increased his overheads and down time.

In this situation of a weakened strategy, as figure 3-4 suggests, meetings are held to refocus the business. The marketing function is required to rationalize the product line; the sales function is called on to restrict its offerings but the price structure is rationalized and generally lowered; the engineering function is ordered to simplify and standardize the product line so that the manufacturing manager can meet his new and reduced unit cost targets.

But the new functions cannot respond in lock step.

The marketing function is a planning staff. It can make its analysis and come up with product descriptions for a revised line in fairly short order. Some commitments may have to be worked out and some advertising or brochures reframed. Redefining the line may be difficult. Some decisions may need to be made that will turn out to have been critical. But the task itself can be accomplished in a few months.

The sales function has more of a problem than marketing, because it is responsible for the actual interaction with cus-

3-5. Time required to change missions.

Business Strategy	Marketing Mission	Sales Mission	Engineering Mission	Manufacturing Mission
Market dominance	fairly short	fairly short	fairly short	very long
Specialty market niche	fairly short	fairly short	fairly long	long
Delivery response	short	short	fairly short	fairly long
Market coverage response	fairly short	fairly short	fairly long	fairly long
Custom product response	fairly short	fairly long	long	long
Product innovation	fairly long	fairly long	very long	very long
Technical innovation	long	fairly long	very long	very long

tomers. Despite the appeal of an aggressive price policy, the new strategy will eliminate some products customers are counting on. The problems are these: making effective substitutions in the customer plant without losing the account; moving obsolete inventory; and servicing products that incorporated the deleted components. The sales function cannot respond quickly to a change because it is paced by the change requirements of its customers.

The engineering function has a major problem. It is never a simple task to standardize a product line, and the whole future of the business will depend on the new line. There must be no design mistakes. In many cases it will take more man-hours to redesign an existing product to reduce unit costs than it took to design the product in the first place. The job may have to be staged product by product, and modular concepts applied that will take effect beyond the needs of the particular products selected for the first stage. It will be a time-consuming operation.

Since the engineering effort has to precede the manufacturing task, the plant will be the last of the functions to respond. Furthermore, effectiveness of the redesign will depend on the nature of the process equipment and perhaps its restructuring or modernization.

In fact, if senior management is geared to the quick changes that the marketing people and the planning system indicate, it will become enormously frustrated by the slow response of the engineering and manufacturing functions in actually doing things that carry out the intent of the new strategy. The level of frustration is proportional to the lack of operational planning that takes place at the time of the strategic redirection (see fig. 3-5).

A Wild Style of Play: The Problem of Detail

Innovative management is a national resource in America, but it causes difficulties that can be avoided without excessive damage to the entrepreneurial psyche. All that is required is a larger dose of reality and that reality is the level of detail required by the different functions in order to implement a business strategy.

Financial strategy deals in round numbers and summary data. The bottom line and the tax effect are principally all that is required.

Marketing strategy is similarly blessed. Its critical factors are an understanding of customer needs, a knowledge of how one is currently positioned in the market, and what resources will be required to move elsewhere.

Deploying the sales function presents not so much a problem of detail as a problem of interfaces: what is required is an intimate knowledge of the decision-making process in customer organizations and the ability to provide some kind of package that meets their needs. It does not require an intimate knowledge of the product anatomy, element by element.

When we turn to the engineering and manufacturing functions, however, the level of detail required to implement a business strategy rises dramatically.

For the engineers, an immense amount of work has to take place to detail the construction of the parts, their specifications and tolerances, and their fit and function as a whole in assembly and disassembly. For the manufacturing people, this detail has to be extrapolated one stage further to cover the flow of incoming materials and the activities of the process operations, backed up by the motions of each one of the workers, who are all governed by detailed operations sheets through to the shipment of the finished product.

In order to carry out their work, both the engineering and the manufacturing people have to obtain information from marketing and sales if they are to define product specifications and size the plant for the long term. Before they can define specifications and tolerances, the design engineers need from the marketing people a precise description of the performance characteristics of the product. This must include marketing's order of priorities, since many of them are bound to conflict.

Such operating details may play themselves out on a level below the planning horizon of the more entrepreneurial general managers. But it is a mismatch in just such detail that creates rifts and chasms that cause entrepreneurial businesses to go bankrupt.

PLAYING THE ROW GAME TO WIN

A few simple points have been made so far regarding the effective implementation of business strategy.

First, the general manager must understand enough about the different functions of the organization to know how to use them to carry out the strategy: he has to know how the pieces move before he can play the game. This does not mean he needs to have been an experienced manufacturing manager before he can direct the business as a whole; it does mean he has to understand at least what the manufacturing and other functions can and cannot do.

Second, it is necessary to establish some kind of lateral communications system to make sure that the interactions between functions are working correctly. Because of the disparity in the time it takes different functions to respond and in the amount of detail they need, certain classic interface difficulties must be recognized, quantified, and resolved.

Third, to test the feasibility of the grand strategy it is useful to have some mechanism for taking a close look at each of the functions in order to provide an early assessment of what is needed to carry out the game plan. For example, one of the functions may require so much time and funding that it is advisable to engage in some less brilliant strategy rather than try to do battle with the particular windmill that has been selected for attack.

In the chapters that follow, the main reference point will be the manufacturing function, because in most industrial corporations that is the one least integrated into the strategic planning process. However, much will apply to the other functions as well; it is impossible to examine the manufacturing function in isolation.

4

The Marketing-Manufacturing Interface

Marketing has been described so far as a small group of movers and shakers whose objectives are carried out by the working crews in sales, engineering, and manufacturing. This is not always true: in some consumer products industries, the advertising and promotion budgets controlled by the marketing staff may be so large that marketing dominates the business. By and large, however, marketing is a planning and steering activity uneasily linked to the three line organizations. Unfortunately the linkage leads to problems.

- Interfunctional balance is not easy to achieve when there is no common language. In order to bring about meaningful communication, it is first necessary to translate the objectives of one function (marketing) into the requirements of another (manufacturing).
- The marketing function sometimes confuses itself with the general business management and generates initiatives that can lead to oversteering or understeering. The initiatives can oversteer the system by dictating commitments that the other functions are unable to carry out; they can understeer when their links into the line functions are insufficient to ensure that the marketing program is implemented.

- The marketing function is itself likely to suffer from turnons and turnoffs. It can shift in fairly short order from a period of low activity to one of high activity when it generates commitments that do not adjust easily to the slower response times of the line functions.

TOO MUCH MARKETING AND NOT ENOUGH

Marketing-driven programs can destroy the rest of the organization unless the linkages among the functions are thought through.

> A Midwest entrepreneur built a $200-million company on a strategy of fast response to new market opportunities for low-priced household products. The organization was geared to fast development, rapid changeovers on the shop floor, and aggressive sales followup. But as fast as his people tried to be in response to his commitments, the owner-president was always one step ahead of them.
>
> As the company grew, the product lines became more complex in design and manufacture. The technology became more demanding, and this slowed down the response capabilities of the organization. But the owner-president, obsessed with the time pressures of possible competitors, drove the organization faster.
>
> On one occasion, tests identified that the product formulation in a new household deodorant could become unstable in use. The engineers suggested a development program to reformulate the product with a proper inhibitor but were overruled because there was no slack in the schedule. Instead, a standard inhibitor used in other products was blended into the compound.
>
> The standard inhibitor caused the product to gel during the manufacturing process and production had to be called to a halt. Not only did a new inhibitor have to be developed while the product was in production, but a new procedure had to be evolved to decontaminate the process equipment.

What is significant in this case is not the misjudgment—that can happen under any circumstance. What is significant is that the owner-president used all the weight of his authority to keep to the marketing proposal he had laid down without coming to terms with the organization's ability to respond. Similar problems arose every year; there was no valid assessment of the implementation requirements and none of the functional managers could stand up to the boss.

This is not to say that a strong marketing mission is bad for a company, but only that it cannot be allowed to run roughshod over the requirements of the other functions. A weak marketing role can cause almost as much trouble as an overly aggressive one. For example:

> A group of young skilled machinists organized a shop to make to order small runs of metal cabinetry for the electronics industry. Their work was of the highest quality and the shop eventually expanded to about 150 workers. While seldom the low bidder, the company won accounts because it had invested in modern machine tools that enabled it to take on the most difficult jobs and do them well.
>
> After more than 30 years in operation, the company had never developed a marketing function. A few salespeople called on the principal customers, but the bulk of the business came from the torrent of bid requests that came in the mail every day. The estimating department went through this material item by item, but there was no analysis of the bidding procedure— no examination of success ratios, no formulation of bid standards, not even an analysis of profitability by type of job. All the owners knew was that at the end of the year they had made money.
>
> Then the senior salesman retired. Meanwhile, a new group of younger purchasing agents had begun to take over in the customer corporations. The owners realized that they were losing touch with their customers. They began to lose bids to competitors located outside their region—and they discovered that half their sales were tied up with a particular customer who was planning to set up his own machine shop.

In this case, the lack of a sense of direction from a marketing function left the company at sea. On the one hand, a new pattern of customer requirements had evolved without their knowledge; on the other, they had developed a mix of production capabilities without knowing how to optimize it. They did not have a clear fix on the market requirements or their own manufacturing capabilities.

OTHER FUNCTIONAL INTERFACES WITH MARKETING

Although the marketing-manufacturing interface is a critical problem area, the interfaces between marketing and sales and marketing and engineering can also lead to manufacturing difficulties.

The Marketing-Sales Interface

In large corporations, most business units are obliged to share a common sales force to some degree. This generates a good deal of drama behind the scenes, for example when the marketing group feels it needs special attention from the sales function to support new-product introduction or to deal with peculiarities in the product line or customer requirements. Since the sales force is compensated on volume of orders, this kind of special treatment is always difficult to ensure, and sometimes it leads to difficulties for the plant.

> A product manager for a specialty line in a major corporation was having difficulty attracting the attention of the main sales force. He campaigned for a small group of specialty salespeople who would be technically trained with the product and responsive to the needs of a growth business. This was a costly undertaking, and the corporation limited him to the one regional district where most of his sales were concentrated.
>
> As a result of the regional campaign, a flood of orders was relayed to the plant, which was entirely unprepared to handle them. Marketing and sales considered the campaign a great success, until they discovered that their plant could not meet the delivery commitments. The orderly scheduling of the plant was upset and costly changeovers became more frequent. Quality problems began to emerge.
>
> The plant manager made a special effort to respond to the new order mix and changed the process control systems as much as he could. However, by the time he was able to respond fully the campaign was terminated because there had been too many broken promises to the customers.

The Marketing-Engineering Interface

The marketing-engineering interface is also a classic stress point. Since the product research budget generally filters down through the organization through a different process than the marketing budgets, the two are not always coordinated programs. Research and development budgets are based on assumptions that cover more uncertainties than the organization may recognize—or wish to recognize—at planning time.

The product manager of a corporation was given two assignments: coordination of a new-product introduction and the general upgrading of the rest of the line.

The research and development laboratories were funded to support both tasks. However, the new-product development program ran into technical difficulties that delayed it, just at the time that the company began to experience a recession in its industry. The chief engineer was replaced and a new strategy was begun to give priority to the completion of the product development program through a different technical approach.

Progress began to be made under the new program. At the same time, however, technical support for the upgrading of the traditional line was cut to the bone. The sales force began to generate pressure for customer service and went directly to the plant to obtain "special effort" for "this one critical customer." By the time the new product was ready to be introduced into the plant, the manufacturing process had become disrupted by a spate of technical changes on the traditional product lines—all of them generated on an ad hoc basis, without any kind of general design direction.

The Marketing-Manufacturing Interface

Examination of the marketing-manufacturing interface has led us to the conclusion that there is a communications barrier between the two that has to be recognized and dealt with, before it will go away.

The marketing staff in a corporation had developed a sophisticated way of defining and reporting its status and objectives. Each product manager positioned his products in market segments that conformed to the growth-share portfolio, identified the competitive trends in each market segment, and defined his market-share objectives. This became a common language among the marketing and planning staffs and much clarified where they were and what they were trying to do.

The system was so effective it was surprising to find the manufacturing managers complaining that they never knew what the marketing people were trying to do. Everything was a secret and a surprise, to hear them tell the story.

When a few of the manufacturing people went over the marketing documents, they were little better off. The marketing people were talking about growth segments and market share, commodity markets and specialty markets—none of which meant a thing to the manufacturing people. The plant people wanted to know about batch sizes, changeovers, frequency of runs, product specifications, and so on. All of this could be pried out of the marketing documents, but it required a translator who understood both functions to do so.

The manufacturing problems generated by marketing can be seen more clearly by focusing on the product management function. Marketing strategy may remain difficult as long as the manufacturing function fails to understand what it entails and is unprepared to cope with it. But it is the management of the product line that constitutes the main fault line at the interfáce.

PRODUCT MANAGEMENT

Several elements determine the actual operating requirements placed on the manufacturing plant: the character of the product mix, its breadth, technology, and specifications; the rate of new-product introduction; the degree of complexity; the rate of growth or decline; and the conditions of the product's performance, together with the customer expectations of them. The marketing people can talk about market segments indefinitely, but it is the product slate that the manufacturing people have to deal with.

Product management varies from organization to organization. In some companies, it is a well-defined middle-management responsibility; in others, it is a nominal function that is actually carried out by the more senior people whenever they feel it is necessary to step in and stir the pot.

Three major plant problems can be handled best by strong product management:

1. Product proliferation—which brings about the gradual inflation of the product line-until it comprises large numbers of marginal products
2. The separation into commodity and specialty segments—which splits the performance criteria of the plant
3. Evolving quality standards—which leads to the moving of performance expectations away from the traditional standards that the plant is trying to meet

Product Proliferation

Product proliferation is so widespread that it seems to be a law of nature. With time, new products will be added to the plant without the deletion of old products and this will gradually increase the plant's disorder level.

Consultants have been making a meal out of this phenomenon for decades, since it is apparently more difficult for people inside the organization to deal with the problem than to call in an outside observer. This is so for two reasons: first, the costs of variety are difficult to identify without special effort because the accounting system is not structured to expose them; second, there are more political constituencies in favor of proliferation than there are against it. The political pressures will be considered first.

> An obsolete product in a manufacturing company kept coming up as a candidate for removal. But it was never removed. Everyone agreed that the product no longer gave value for money, either for the customers or for the company. It was difficult to make, ran infrequently, required a high inventory for decreasing sales, and interrupted the rest of the production scheduling.
>
> Whenever an attempt was made to remove it from the line, however, the Denver district sales manager raised a storm. He had a major customer who depended on that product; no substitute was acceptable, and any attempt to close out the item would lead to a long period of testing during which competing products across the whole range of the product line would be brought in for evaluation. As long as that product stayed on the qualified list, it had better not be withdrawn or the whole line would come into question.
>
> The marketing and manufacturing people were just beginning to rebut this argument when the cost accountant added her story: "Well, you can take that one out of the line if you want, but it won't lower your costs. We'll just allocate the burden over the rest of the line."

The best way of dealing with this kind of problem is not to fight it out item by item. A better way is to work out some general agreement among the functional managers as to how wide the line should be, and then let sales and marketing decide which products have to be dropped. It becomes easier for them to argue that issue when they know they have to produce a list whose length has already been agreed on in advance.

How does one decide how wide the line should be? By working the trade-offs between marketing and manufacturing.

If manufacturing can establish that this batch runs longer than such-and-such it can reduce costs to X and if marketing testifies that if it has to drop the smaller-volume products it will lose sales of Y, there is one wing of the trade-off. On the other hand, if marketing can increase its product width to Z, it will contribute such-and-such additional sales volume at an incremental cost estimated by the manufacturing people, and there is the other wing of the trade-off.

General management makes the fundamental decision based on interfunctional estimates of costs and opportunities, and the marketing and sales people are locked up in a room until they produce enough answers to make it happen.

Separation into Commodity and Specialty Segments

As a product line matures, it tends to split into a price-sensitive commodity portion and a performance-sensitive specialty portion. This is a classic issue for the marketing people to deal with. Should they slug it out on price to maintain market share? Should they focus attention on the specialty markets, emphasizing design, product performance, wide variety, and responsiveness? Or should they try to do both, segmenting the two tasks and dealing with them separately?

> I have sat in on a number of such discussions and found the issues treated with perception. I wish I could report that the perceptiveness led to an effective response. I can't do that, for the reason that most of the discussions took place without the participation of the manufacturing people, who were the only ones in the whole organization that could have shed some light on the topic. Instead, we usually had the benefit of counsel from cost accounting.

The problem of dealing with these issues from a cost-accounting point of view is that the range is necessarily limited. The controllers can say how costs will behave if more or less unit volume is loaded onto the plant or if the product mix varies a little. But they cannot say what the cost structure of a plant dedicated to commodity product would look like because they do not know. Nor do they know how a plant designed for a mission of product responsiveness in the specialty market would be configured.

These are manufacturing questions, though they usually have to be translated into manufacturing language before the plant people can understand them and respond. If manufacturing people could anticipate a commodity product slate and were able to push the plant up the scale to a 7 level, they should be able to say what they could do to change the present cost structure.

In the plant, the issue comes down to the minimum market necessary to keep separate missions facilitized. If the marketing people can "guarantee" a unit volume of no less than such-and-such for so many years, the manufacturing people can set up a dedicated line that will require a certain level of capital funding and will produce at such-and-such a cost level. If the marketing people can define what a specialty product mix will consist of, manufacturing will be able to provide responsiveness of such-and-such a pattern.

In other words, the manufacturing response will come in the form of more issues: if you provide us with funding of *A* and a product slate of *B,* we will be able to meet criteria *C.* This requires a good deal of dialogue to resolve, but eventually the alternatives can be understood, the underlying assumptions rationalized, and the problem focused for decision making. Even a decision to keep to a specialty product line need not leave the manufacturing people helpless. If they know they will have to work with a plant at the 3 level, they can at least produce plans to make it work effectively under that level's process, control, and product conditions.

Evolving Quality Standards

A major problem emerged during the 1970s. It became clear that customers, in part tutored by imported products, had established quality expectations that went considerably beyond the traditional manufacturing standards in American industry. One company after another woke up to the issue long after it had become a problem. If management is the anticipation of problems, then there was little in the way of management being practiced in this country.

> A company making subassemblies for consumer appliance manufacturers was abruptly introduced to the issue during the 1980–82 recession when shipments their customers had formerly accepted were suddenly rejected. The plant was subjected to an invasion of field inspectors making source checks of its quality assurance system.
>
> The plant managers could not understand the problem. There had been no changes in the labor force, the work standards, or the inspection procedures. Moreover, there was no pattern to the rejections: some customers complimented the plant for work rejected by others. Even within a single shipment, some items were singled out for criticism that appeared to be no different from others that were accepted.
>
> The received wisdom at the top of the organization was that all of these problems were traceable to the economic recession, and that they would go away when it did. Customers were being choosy, they thought, because they could afford to be, so there was no real need to take drastic action.
>
> The salespeople, however, reported that they were dealing with a new generation of purchasing agents: companies looking for suppliers who could meet more precise tolerances. They were trying to upgrade their own products in order to meet foreign competition, and they were serious about finding suppliers who would support them.

In fact, many industry trends were leading to a major reevaluation of quality assurance systems. It was not a short-term phenomenon connected with a recession; a major rebalancing of product performance priorities was being called for. The reasons are discussed below.

Value Analysis: In the 1950s products were designed with some performance margin: sheet metal was heavier than necessary, thick coatings were applied, and there were plenty of rivets and screws and bolts to fasten the product together. But in the 1960s new materials and metal-processing methods appeared and the product designers began to take them seriously. If a coating maker claimed that one mil was all that was required, there was no need to specify three; if a structural adhesive manufacturer claimed it could replace rivets, then product design could eliminate all the metal fasteners. Soon the suppliers of materials and components found their products exposed to their limits. They could probably meet customer requirements, but there was no longer any margin for error.

Stressed Performance: Many products were used in increasingly stressed performance conditions. When oil drilling operated at 5,000 or 10,000 feet, the blowout preventers had one set of quality standards; when depth reached 25,000 feet, an entirely different set of performance criteria had to be met because the temperature and pressure conditions became severe in more than a linear way. This kind of problem came up in the military and space programs, medical and scientific instrumentation, aircraft and automotive performance, and even in household appliances and toys. Sometimes the problem arose because the product was exported to more distant markets and very different climates than previously.

> A luxury automobile designed for road conditions in the United Kingdom and Western Europe was being heavily exported to the semiarctic conditions of Canada and Scandinavia and to the tropical conditions of the Middle East and elsewhere. In Phoenix, Arizona, the dealer regularly took out the company's magneto system and substituted an American make designed for extreme temperature conditions. The home office had a difficult time believing that any climate would be so severe as to require this kind of substitution.

Multiple Interaction: If a product has few moving parts, the probability of failure is limited; when the number of parts increases, the probability of failure is greater than the life expectancy of any one of the components. They interact with each other, and the chances of failure increase due to the sheer number of things that can go wrong. Product quality has to be significantly better in a contemporary house with more than a dozen electrical appliances and no available service than in a house of an earlier age with few appliances and plenty of available service.

The increase in product complexity is notable with automobiles and aircraft, which require far higher standards of reliability than in the past just to equal the service life of the older products. It is a major problem in the armed services, where the reliability record of high-technology weapons systems worsens as the number and complexity of their components grow.

Competitive Performance: Consumer quality expectations have risen simply because customers discovered the desired finish and reliability could be found in foreign products. This has been the case with one industry after another, from consumer electronics and appliances to automobiles, where first German and then Japanese companies set a standard that American plants were not prepared to meet.

Product Liability: At the same time, product quality became a high-risk issue, not only because of the danger of customer alienation but also because of the real danger of negative court judgments. Contaminants and failures, toxicity and fire hazards could lead to punitive liability settlements that would bankrupt a manufacturing company.

The Key to the Solution

The key to the solution of the issues created by evolving quality standards is to work the product specifications. "Working" them means involving all the functions in a joint effort so that specifications can be established that are sound from the perspective of both customer use and sound design, and that are producible.

American industry eventually took these issues seriously and has been turning the whole system inside out in order to tackle the problem of quality assurance. But it is a difficult battle and still goes on. The following example is indicative.

> During the annual planning cycle, the issue of product quality came up and received the usual reassurances from the manufacturing managers. All the regular systems were being strictly enforced, they said.
>
> Out of curiosity, one of the senior managers inquired what these regular systems might consist of. She learned that the plant was unable to meet most of the product specifications, so it tested the product and set aside the off-grade material. The off-grade was then blended into another batch that happened to be particularly splendid so that the mixed product would just meet the minimum specifications most of the time.
>
> This description of a regular system put the senior managers into something close to shock. They wanted to know why the plant could not do better. By way of explanation the manufacturing people cited one capital funding proposal after another that had been turned down by headquarters.
>
> That is not the end of the story. The manufacturing people were unable to guarantee they could produce to specifications first time around even if the new equipment had been authorized. Their proposals had been for general modernization projects. The embarrassing fact was that no one

knew what it would require to meet specifications. They had worked the present system so long that they had no technical data to help them bring the process under control.

In this case as in others, when senior management finally took the quality issue seriously, they discovered that the problem needed a lot of homework before a viable strategy could be developed to improve the situation.

BARRIERS TO THE MARKETING-MANUFACTURING INTERFACE

In order to meet the requirements of the marketing function, whether it is a matter of supporting the long-term strategy or the operating interface of the product management, the manufacturing managers have got to get into direct dialogue with the marketing managers and reach an agreement as to what can and cannot be done. To do this, some limitations and hazards that have become built into the organizational structure over the years have to be overcome. The principal ones are these:

- Limitations in the range of responsibility accepted by the manufacturing managers
- Plants that have been buffered from direct contact with other functions
- Communication systems that fail to maintain an adequate flow of information
- Plants with minimum standards or with ingrained objectives that work against the needs of the other functions

Range of Responsibility

The manufacturing manager in some industrial companies still thinks his job covers the plant, from the receiving dock to the shipping platform, and that is all. This is important, to be sure, but it is only the most obvious portion of the manufacturing function. Until manufacturing managers think in terms of a satisfactorily delivered product, they are going to suffer from their tunnel vision and everyone else will suffer as well.

The manufacturing managers proper range of responsibility will reach into the product specifications, the purchased materials and components, the shipping system, and the full range of manufacturing services—everything that is necessary to make and deliver acceptable product.

Buffered Plants

In some organizations the plants are buffered from the marketing and sales activities, and of course from customers, perhaps on the assumption that it might confuse the manufacturing people to know what was going on in the field.

> An elaborate operations department had evolved in one corporation to bridge the gap between the plants and the marketing and sales activities. The operations people scheduled the plants, worked up the plant modernization plans, and interfaced with the sales and the marketing people. Any plant contact with the outside world was carried on through them. The system virtually isolated plant managers from the rest of the organization.
>
> Eventually the group vice president and the senior manufacturing manager decided they had had enough of it. They reorganized the company into strategic business units, placed a general manager in charge of each, and invited the manufacturing managers at plant level to become directly involved with the planning of the business teams. The buffers came down and the manufacturing people received a welcomed opportunity to work out their problems directly with marketing and the other members of the team.

Faulty Communications

When the manufacturing managers see the need to establish communications outside the manufacturing function, one thing commonly happens: they set up a system, but it does not work.

Take as an example customer feedback with regard to quality problems. It would seem simple to obtain information about product quality. In fact, it is nothing of the kind.

There is something called a "bathtub curve," which traces the path of product difficulties over time. The curve is high in the beginning due to the various bugs and installation problems common to the introductory phase of a new product. Then it settles down to a period of maintenance-free operation, until finally old age catches up with it and the incidence of difficulties goes up again.

In order to catch the introductory problem pattern, most manufacturers enclose warranty cards with the product and ask customers to fill in the date of purchase and report any problems. Few do so, and for those few the correlation between a true date of purchase and the date a warranty card is filled in is not high. The complicated feedback systems used with more sophisticated installations also cannot be relied upon to work without careful supervision.

> A computer corporation set up a staging area to bring together units manufactured at different plants in order to provide the complete systems ordered by their customers. The area was intended to be no more than a depot, without a test facility or other technical support. The real test of the system as a whole took place on the customers' premises, often in remote locations.
>
> To make sure that the several products were compatible as a system, the installation teams were required to fill out reports describing problems, and these were then forwarded to the plants.
>
> A check of the field reports against some actual installation visits indicated that only a fraction of the real problems were getting through to the plant.
>
> It turned out that the installers were compensated on the basis of the number of hours required to get the system working. One pattern of difficulty paid off well, and they tended to filter the reports toward it. The plant managers already knew about those problems, however, and believed they had taken care of them. Meanwhile, the real difficulties went largely unreported.

Plants with Limited or Traditional Objectives

Not all manufacturing managers recognize the need to respond to marketing requirements or to take the new quality standards seriously. They continue to work with the obvious tasks of output and unit cost without trying to deal with the longer-term requirements or the hidden costs.

> The traditional objectives of a manufacturer of automotive components had been to meet the volume requirements at the lowest cost of production. Management had positioned the company in the industry by negotiating

low prices and delivering the product on time, and that is the theme the plant worked to.

When statistical quality control requirements started coming through the industry, the managers did not take them seriously. Their components were delivered to a subassembly plant rather than directly to the automobile companies, and they thought they were too far back in the tier of supply to require any significant changes.

The company managers were forced to come to terms with the problem of quality assurance when their supply contract came up for renewal. They were put on probation for a year; one of their customers warned that he was looking for an alternative source of supply that could guarantee quality of the components.

The senior management became alarmed and put pressure on the manufacturing managers. A quality circle was started, a new quality assurance manager was hired, and some new machines were installed to improve control over the process.

But six months later it was difficult to detect any change of attitude around the plant. The quality assurance manager complained that he got no support from the plant manager. The quality circle was soldiering along on a few picky items. And when the issue of quality was raised with the foremen, the answer was something along the lines of: "Sure, we'll do the best we can. We're a lot better than we used to be. But don't forget, we still got to get the product out the door."

THE COST OF VARIETY

Manufacturing people know intuitively that some marketing strategies are going to disrupt the plant and build up the costs, but they have a difficult time quantifying what the costs will be.

In general, the cost of variety is greater than the accounting system reports it to be. Burden allocations only serve to cover it up; the slow drift into product proliferation masks it. Yet the cost of variety is probably the most significant element in the overall cost structure of the plant—particularly in those plants that do not recognize it or organize themselves to deal with it.

Consider the difference between making a wide line of short-run and changing products and a limited line of stable ones. The former creates burdens throughout the company:

- The purchasing department, the materials management, the vendor control and assessment, the incoming inspection, and the stock of incoming materials

- The production lines, which cannot get up to speed to establish learning curve rhythms or obtain dedicated machinery to optimze the runs
- The quality-assurance function, which must develop separate standards and procedures for different processes and products; this results in a quality-control risk, particularly to the minor products that are run infrequently or that require special expertise
- The increased investment required for both process and finished inventory
- The warehousing operations and the distribution system, which have to stock many items, replenish them, and meet the production scheduling and logistical requirements they impose
- The administrative functions and systems support, which must develop complex and detailed accounting procedures to track every separate batch

These costs are not traceable to any one product; they are due to the total effect of proliferation across the product line. How does a company determine what they are?

They can be estimated by determining what kinds of efficiencies the plant could obtain if it pursued a strategy of rigorous standardization and maximum unit volume output from a limited product line. It is a what-would-happen-if exercise, not workable through the existing accounting system or traceable to specific products but capable of providing a base case against which to measure the cost of variety.

A company making electric motors for a variety of specialized applications had allowed the product line to proliferate over a period of time. To differentiate the company from the competition, the sales force had developed the habit of offering to customize the product for industrial customers even though equivalent standard motors might have met the requirements. Since the engineers were never provided with overall guidelines, each design was unique. Analysis of the product dimensions showed a range of motors in the band from 4 13/16 to 4 15/16 inches, when a single 4 7/8 housing would have sufficed.

The result of this proliferation was that the manufacturing plant was obliged to use general-purpose machinery. It planned and scheduled production in small batch runs and spent a considerable amount of time and effort stopping and starting machinery to get it ready for the next lot.

A single high-volume order enabled the plant to set up a dedicated line in one corner. In an attempt to find the cost of variety, the performance of this line was applied to a theoretical plant operation where the product line had been standardized around a limited number of configurations and dimensions, reducing the number of products from 132 to 17.

The study indicated that significant savings could be made in direct labor, materials, indirect costs, materials handling, and inventory investment. One of the surprises was the effect on response time through the plant: instead of a six-month lead time to put a product through the system, the company would be able to respond in as many weeks.

As high as the cost of variety is—and a figure of 30 percent is not unusual—there is an associated cost caused by management uncertainty. Decisions regarding the width of the product line are delayed because the marketing people want to keep their options open and general managers want to see how the economy will move. But manufacturing managers need stability to work their operations on a cost-effective basis. Someone must make a commitment and let the plant get to work. The assumption may turn out to be wrong, but at least the plant will not be switched on and off to meet the latest data.

The appendix consists of a paper by William L. Wallace entitled "A Different Way of Thinking About Manufacturing Costs." It considers five kinds of costs in manufacturing.

1. The inherent cost of manufacturing typical products as designed at the volume level of the total throughput for this general category of product
2. The cost of variety that arises because the manufacturer makes more than a single product, either by producing a number of different products or by producing a number of variations to achieve a greater breadth of product line
3. The cost of change, including the cost of new and altered products, engineering changes, and the cost of scheduling changes made to meet unpredictable or rapidly increasing demand or to deal with substantial seasonal or cyclical variations in volume

4. The cost of system and organizational suboptimization, which can arise from a number of factors, such as suboptimized inbound and outbound freight, over- or undermanning, over- or undersizing of facilities, internal frictions within the organization, and so on
5. The costs of writing off past decisions, worth segregating for decision-making purposes because they deal with the issue of how to allocate costs that have already been incurred

Since these considerations go beyond the mission of the regular accounting system, they have to be examined without the benefit of the existing reporting structure. Some will be estimates, but at least the company will have a way of assessing the cost of variety as a whole rather than having to fight a losing battle over each marginal product one at a time.

MANUFACTURING TO MEET LEGITIMATE MARKETING REQUIREMENTS

Seven problem areas have been identified in the interface between marketing and manufacturing.

1. Overstressed and understressed marketing initiatives not integrated with the other functions
2. Marketing strategies that have not been translated into terms meaningful to the manufacturing manager
3. Coordination problems between marketing and the other functions that lead to erratic order patterns at the plant
4. Product proliferation, leading to a high cost of variety in the manufacturing function
5. Unresolved competition for manufacturing resources between the commodity and the specialty product lines
6. Manufacturing managers who have not kept up with the evolution of customer expectations with regard to better quality assurance
7. Management inability to make decisions under uncertainty, causing the plant to operate under unstable and changing conditions

There is an eighth issue, which was mentioned in chapter 1. New process technology exists that enables the manufacturing function to operate cost-effectively with greater variety and shorter lead times than was possible two decades ago. This will be discussed later, but it should be noted here that the optimal management is a two-way agreement between the functions.

On the one hand, marketing and sales should understand the kinds of requirements that are going to stress the plant beyond its design limits. On the other hand, manufacturing should keep sufficiently current with the state of the art to know what can be done with modern process technology in order to support changing or varied requirements.

5

The Sales-Manufacturing Interface

In the past 20 years a number of American industrial firms have created logistical nightmares for themselves by encouraging delivery commitments without ensuring the support necessary to sustain them. Sometimes, under favorable conditions, the system works. But as product lines proliferate and market conditions change, it has required increasing quantities of inventory, and this has led to imbalances, delays, and excessive capital investment.

In part, the problem is inherent in America's geography and economic diversity. This is not Japan, where both major population centers and industrial sites are located along a single rail line. The United States has a complex and farflung array of vendors, manufacturing plants, warehouses, distributors, and consumer outlets. They have seldom been organized to respond with the flexibility needed to meet the commitments made on their behalf.

PROBLEMS WITH THE TRADITIONAL SUPPLY SYSTEM

The traditional supply system can be thought of as a three-part interaction. First, the sales force responding to individual customers; second, the logistical system trying to support the order pattern; third, the management override monitoring performance and adjusting the system to meet financial targets.

The Sales Force

In a traditional system supported by stocks of finished inventory, the principal difficulty for the sales force is not the pattern of orders. Rather it is the unanticipated departures from the order pattern it has established.

A logistical system can be organized to respond to virtually any order pattern and it works if the pattern remains relatively unchanged. In fact, if the order pattern behaves itself over an extended period of time the system tends to become optimized. Marginal orders are questioned and reduced, lead times are adjusted to replenish stocks, and response times are generally understood and adhered to. The system that evolves may not be efficient—some elements may be loosely operated while others are harnessed under strain and a lot of capital may be required to bankroll the inventory—but it can meet its commitments. The problems arise when the order pattern moves beyond the band width the system is set up to respond to without allowing the time necessary for the system to adjust.

Unfortunately, we do not live in a well-behaved economy. Whole industries are subject to boom and bust; export campaigns and import penetrations unbalance the economy; changes in interest rates pull in new capital investment or dry it up; technological breakthroughs and market fashions render entire markets obsolete; new products are introduced and succeed or fail; product liability suits and other calamities wipe out established companies. Even slow drift can accumulate and shift an order pattern outside its supportable band width.

Consider what takes place when an out-of-band demand is linked to a slow response system. Look at figure 5-1. The vendors and the manufacturing plants are in step in January with sales demand at a rate of 100 units per month and stocks of raw materials and finished products at 40 units each. Now, the demand increases by ten units each month for five months between March and July and then decreases again to December.

5-1. Supply to a well-behaved pattern of orders.

Month	Vendors	Materials Inventory	Production	Finished Inventory	Demand
Jan	100	40	100	40	100
Feb	100	40	100	40	100

By March and April (fig. 5-2) the increased demand has depleted the stock of finished product but has not affected the plant. By May the company has to start backordering and the plant schedulers are predicting further increases and calling for a raise in the output rate. But the plant cannot increse its production very fast, so it falls farther behind the sales orders. By July the plant is on overtime. But it has run out of raw materials and has to search for emergency sources of supply while calling on its vendors to increase their shipments.

In figure 5-3, the order levels are decreasing again, back to the original level of 100 units. But there is pandemonium in the plant. In August it is still behind, so a second shift is put on and production goes up to 200 units per month. This has the effect of building up the finished stocks so high that they are cut back in November. But they are still too high, and in a declining market the plant is instructed to reduce operations in order to get back to the original level of 40 units of finished product in inventory. By now the vendors have geared up to 200 units a month and the raw material is flowing into the plant just when the demand has dropped—and may continue to drop below its 100-unit level to compensate for the increase during the year.

5-2. Supply to an increasing pattern of orders.

Month	Vendors	Materials Inventory	Production	Finished Inventory	Demand
Mar	100	40	100	30	110
Apr	100	40	100	10	120
May	100	35	105	(15)	130
Jun	100	20	115	(40)	140
Jul	105	(5)	130	(60)	150

5-3.   Supply to an amplitude variation.

Month	Vendors	Materials Inventory	Production	Finished Inventory	Demand
Aug	150	(15)	160	(40)	140
Sep	175	(40)	200	30	130
Oct	200	(40)	200	110	120
Nov	200	60	100	100	110
Dec	—	20	40	40	100

These basic concepts in industrial dynamics were worked out by Jay Forrester as far back as the 1950s, and the events have occurred with real plants many times since then. What happens is analogous to the amplification of a radio signal. An amplitude variation takes place in the order pattern; the logistical system delays the signal and amplifies it, which causes the greatest pressure on the function that can least respond—the manufacturing plant. Here, a swing of 50 percent up and down has been transmitted to the plant as a demand for 100 percent increase, only to be cancelled by the time it is able to respond.

The Logistical System

How do the logistics work in a traditional supply-demand response system? The stocks of materials in the system buffer the signals between the demand and the production. This may be a good thing for the purpose of absorbing minor fluctuations within the established band; but any significant demand beyond that range causes the system to oscillate out of control. Delays in communications only accentuate the impact of a demand when it is finally transmitted to the plant. Unless the company has a listening post inside the end-user market, it will be torn apart by demands for immediate response for which it is not prepared.

A manufacturing plant is not a logistical system; it is a mere waystation in the long line of supply from the materials producers to the makers of component parts and subassemblies on one side, and the warehouses, distribution points, and consumer outlets on the other. All of this is linked by truck, rail, barge, or air transport, and processed through an elaborate system of purchase orders, invoices, and shipping documents.

Nor is the plant itself likely to have a single comprehensive control center to mastermind its requirements. It is more likely to have a rather creaky administrative apparatus that has been pieced together by a number of different systems managers at different times. It is usually still under construction, sometimes without the benefit of an architect, made up of segments of systems designed for local purposes by people who have since gone on to some other assignment.

There is talk of a date in the future when every fragment of information, from the bill of materials to the operations sheets, from purchases to deliveries, will be organized into one comprehensive computer data base out of which the logistical system will operate as responsively as necessary. So far that magic date has been a moving target. It is usually defined in terms of numbers of years from time zero.

Management Override

In addition to the sales force and the logistical system, settled in behind all this is a kind of volcano in the form of a senior manager looking for return on net assets. The volcano is sensitive to last month's performance figures, and there is a fault line connecting it to the summary inventory data. Whenever the inventory turnover slows down to a critical figure, which is different for every volcano, an ominous rumbling noise will be heard followed by an eruption. In the late 1970s the increasing cost of capital widened the fault lines and magnified the outbursts. A number of industrial companies issued a general decree to reduce stocks by some percentage by such and such a date.

These overrides from above generally have the effect of breaking up the system. When called upon to reduce inventory by a deadline, the plant does only what it can do. It cannot reduce the inventory of slow-moving or obsolescent stock because nobody wants to buy it; so the plant cuts back production and reduces the inventory of the faster-moving items.

But as soon as the plant starts scheduling for inventory reduction it is no longer able to schedule production to respond to changes in demand. When demand goes back up, the plant is off balance: it is operating below its optimum level and has to be revved up again. The inventory is also out of balance, and the production schedule has to regain volume to restore its stock position and rebalance inventory just as out-of-stock items have to be inserted into the schedule to meet critical orders.

You can imagine what happens to production costs when the production scheduling is pulled around by the nose like this.

POTENTIAL OF THE NEW SYSTEMS

The new approach to the organization of a responsive logistical system is based on four elements:

1. A design that defines the overall capabilities of the system and establishes objectives for all the parties involved in it
2. An agreed-upon set of service levels and guidelines for the sales department that will enable them to determine what commitments they can make
3. A strategy for the manufacturing function that will allow it to develop the degree of responsiveness required by the systems design
4. A single comprehensive language, such as MRP-II, to establish communications among the functions and enable management to guide it toward the needs of the business

Systems Design

If it is to establish a viable systems design, the business needs two kinds of information. The first kind comes from the sales function.

- Statistical distributions or order sizes, by stockkeeping unit
- Statistical distributions of order timing, by stockkeeping unit
- Demand trends in the number of stockkeeping units and forecasts of changes in order size and timing
- Lead time requirements to meet customer delivery requirements at different service levels

- An assessment of the incremental market value of higher ser-vice levels
- The degree of uncertainty in the forecasts, with probability distributions to establish a band width about the forecasts

The purpose of the data base is to establish an agreed-upon baseline from which to forecast demand, and hence to discuss the value of alternative response strategies. A conservative re-sponse will hold the bulk of the business but will not attempt to use responsiveness as a competitive tool. An aggressive re-sponse strategy may not try to provide the ultimate service for every product in the line, but it will certainly impose new kinds of flexible response requirements on the manufacturing function.

A Conservative Response Strategy

- The number of stockkeeping units will be limited by elimi-nating marginal products and placing a lower limit on the size order accepted.
- Lead times provided will be consistent with accepted practice in the industry.
- A service level of 95 percent will be aimed at for the main line of products and of 90 percent for the lower-volume items.
- Product development activity will be constrained so as to standardize the line.
- The band width of responses will be limited to two sigmas of the forecasted line.
- Inventory will be minimized by stocking only the main line of products and making the specialty items to order.

An Aggressive Response Strategy

- The number of stockkeeping units will be widened to whatever is necessary to keep pace with new market potential from small or specialty customers.
- Lead times will be faster than general industry practice, in-cluding overnight delivery if necessary.

- A service level of 98 percent will be aimed at across the product line.
- New products will be introduced wherever necessary to update the line.
- Response to immediate market opportunites that may be over and above the band width of the forecast will be the best manageable.

Defining both strategies supplies two points on a curve that will help management assess the incremental value of increased responsiveness. The sales and marketing people may be needed to assess the incremental sales return from the more aggressive strategy, but the gaming inherent in the procedure can be reduced by linking their remuneration to the results they forecast.

The second kind of information the business needs comes from the manufacturing function.

- Raw materials lead times, with probability distributions of the major materials and components and a list of critical items
- Process stages and lead times
- Opportunities in the process technology to reduce lead times by faster changeover procedures, together with estimates of the capital investment necessary to follow them up
- Capability of the control systems to manage an aggressive response strategy for a wide line of products

A conservative response strategy from the manufacturing managers will do everything possible to reduce the response time without requiring significant new capital investment, but its major objective is to reduce manufacturing costs. An aggressive response strategy adds to the cost and difficulties of the manufacturing managers. It is necessary to know how much incremental cost will be involved and at what point on the service curve the costs will begin to rise steeply.

A Conservative Manufacturing Response

- A high-speed flow line will be set up that is designed for maximum output of a limited range of products and that will provide automated stations wherever necessary to reduce unit costs or maintain the output of the line.

- Quality controls will be built into the system wherever possible, either through automation or product design.
- A simple materials control system will be established that ensures supply and throughput.
- Vendors will be organized to meet the business's requirements with minimum buffer stock between them.

An Aggressive Manufacturing Response

- Flexible machine centers that are highly programmable will be provided wherever necessary to process the wide range of products manufactured.
- Workers will be encouraged to participate in quality circles and other activities that will make them responsible for the output of a changing product mix.
- Computer support will be established so that the requirements of a changing schedule can be met.
- Management will work with the design engineers to make sure that new designs are as compatible as possible with existing facilities and practices.

A number of other considerations also affect a viable systems design. Again, they are represented by two points on a curve. The objective is to assess the optimum level between increased manufacturing cost to support an aggressive response strategy and increased contribution from the larger sales income. In order to establish this design objective, the additional factors to be considered include those that follow.

1. *Stability of the product line:* Are any major technological or market changes anticipated that will affect the product line?
2. *Predictability of the product requirements:* Will the degree of uncertainty swing the product requirements frequently beyond the probable band width? How reliable has forecasting been in the past?

3. *Value of improved information:* How critical are certain kinds of data? Is it possible to improve the information base to cover those items? What will it cost to do so?
4. *Financial guidelines:* Do the margins justify an aggressive response strategy? What limits should be established to capital investment in the business?
5. *Human factors:* What are the implications of the alternative strategies for the people involved, both in work-force reductions and in training and operator skills?
6. *Management policies:* What kind of management guidelines should be built into the system? How frequently should they be reviewed?

Role of the Sales Function

The sales managers will have two difficulties with an aggressive response strategy: defining it and controlling it. Considerable effort is required just to define what a company's service level is to be. It is also not an easy task to make sure that the salespeople in the field are making commitments that lie within the definition.

The customer-service level includes the following parameters:

- General objectives in terms of the acceptable number of stockouts or the percentage of on-time delivery
- Delivery objectives by order classification, in terms of type of product, frequency of order, size of order, and location
- Normal lead times from receipt of order to delivery, and priority lead times acceptable for a stated percentage of the total
- Order-inquiry status reporting and handling of samples and special orders
- Processing of special information such as credit inquiries, type of packaging, and marking
- Acceptability of early, late, and partial shipments

A given delivery date can mean a number of different things:

- The customer was vague about when he needed the product because he had some material in stock, so the salesperson put down a nominal date.
- The customer needed the material immediately, but the salesperson put down a nominal date to cover the normal lead time.
- The customer does not really need the material at that date, but the salesperson wants to impress him with a fast response.
- The date on the order is a "drop dead" date, which if missed will hold up the customer's entire factory.
- The customer wants special service, but he is resigned to a "best efforts" response.
- The order is part of a regular monthly shipment, and there is a range of acceptable dates for delivery depending on what is already in the pipeline.

A more precise language is needed to communicate these alternatives and program them into the system. Code designations for this purpose have been developed by some companies. (see fig. 5-4). Such codes perform several functions. They can: provide a data base to define the present service level; help identify where greater response is needed; assess delivery performance; help identify produce codes or salespeople who overload the system with unneeded priorities.

5-4.   Delivery codes.

01   Normal date within regular lead time of a product that is kept in inventory

02   Special priority delivery of production order

03   Customer must have order as specified on precise date, or be notified in advance

04   Customer will accept partial shipment or late delivery if notified in advance

05   Special or sample order, ship by fastest means regardless of freight cost

Without such a framework, a strategy of responsiveness is difficult to enforce because everyone within the system can interpret a demand differently.

IMPLICATIONS FOR THE MANUFACTURING FUNCTION

There are three different approaches to improving the service level.

1. *The traditional approach* meets the requirements through an increase in inventory. When the need to improve the service level is mentioned, the manufacturing people will respond by figuring out how much more inventory is needed.
2. *The modern approach* remodels both the plant process technology and the control system. These managers will propose a plan requiring complete modernization of the plant and its materials management activities.
3. *The sober approach,* which requires a good deal of effort and discussion, adapts existing processes to improve changeover technology and to organize the control activities in such a way that inventory investment is minimized without an elaborate control system.

The Traditional Approach

Improving service by an inventory buildup can be a sensible way to meet a temporary increase in demand, but it is not valid for the long term. Generally speaking, it is no longer cost-effective to buy responsiveness through a buildup in inventory. With the increased cost of capital and the rapid changes in customer requirements, there is a real penalty for carrying this policy too far. The whole manufacturing function must be made more responsive to change, not buried under piles of finished product.

The Modern Approach

The modern approach overwhelms the problem with technology. The three principal ways of doing this are intended to link together and form a single comprehensive whole. They consist of materials requirements systems, flexible machine systems, and computer-integrated manufacturing systems.

Materials Requirements Systems

Materials requirements systems impose a fundamental discipline on the manufacturing and engineering functions through their requirement of a thorough definition of all the purchased or manufactured parts and materials needed for the assembly of the final product. Duplicates are eliminated, where the same part is purchased under different code numbers because it originated from different parts of the engineering organization. A complete bill of materials is needed before new products can be purchased. Some advance commitment must be made as to when purchased parts will be needed and what specifications they must meet when they are delivered.

In other words, even the preparatory work required for a materials requirements system cleans up the bills of materials and the product codings so that component parts for all assembled products can be managed across the board.

When this is accomplished and the coding established, two technologies can then handle the materials. Computers identify parts requirements by exploding the assembly schedules into individual components and materials by date required; and automated or semiautomated picking systems marshal materials and kit them for assembly. Where the product line is complex and changing, the task of preparing these systems is formidable.

Flexible Machine Systems

In addition to the technology developed to computerize the materials control systems, process machinery in almost every industry has been made more flexible than in the past. In the area of basic machine tools, for example, programmable machine centers do the work of a whole shopfull of lathes, milling machines, and drill presses, without manual setup other than the positioning of the part and the off-line programming of the tape controls. In stamping machines, programmable Wiedematics supplant lines of punches and presses. In process equipment, compensating control mechanisms adapt to changing raw materials or drifting parameters by changing the controls while the process is taking place. In welding and painting and light assembly, programmable robots mimic the activity of a skilled worker and conduct the operation without faltering.

All this technical advancement has been brought about by hundreds of design teams working with thousands of equipment suppliers, all working to resolve particular problems or improve specific pieces of machinery. The result is a form of anarchy. While we have in the computer a technology that can integrate everything in the plant that can be made to interface with the system, we have in the machine processes a series of individual technologies that don't interface with much of anything at all.

Computer-integrated Manufacturing Systems

The modernists hope one day to integrate everything that has to do with manufacturing into a single computerized system. Production requirements will be worked back through to the materials to make sure that everything is available on time or to identify the missing parts so that the schedule can be revised; the flow of information from engineering to manufacturing will be standardized so that computer-aided designs can be directly translated into computer-aided manufacturing instructions that program the machines; the system will in effect keep the books on materials purchased, invoiced, released, and used; it will schedule and reschedule to optimize facilities, manpower, and whatever else management wants to optimize.

However, the more complex these systems become, the more disciplines they impose on the operation. In a fully integrated control system, scheduling and purchasing of materials are based on expected lead times. Once it is known how long a product takes to assemble and when it is wanted, a starting date can be defined; once it is known how long it will take suppliers to deliver, a true delivery date can be set. But in any manufacturing operation changes are continual—to revise schedule priorities, alter specifications, make design changes, substitute materials, change the product mix. The system has to be constantly updated.

The great many interactions among the parts of a large and complex system make such updating extremely difficult. More, the entire system has to be tailored to the needs of each particular organization. Man-machine interfaces have to be allowed for or the system will become brittle; yet as soon as that is done, human error and conflicting human judgment become important elements in the system's operation.

Thus far the problem has not been solved in an elegant way. Industry is still feeling its way forward to determine how much complexity can be taken on, how much judgment to allow, and what decision rules to apply to the system.

The systems approach has been applied successfully to simple processes and limited product lines. But there is considerable unfinished business remaining if this workhorse is to be tamed and trained for most manufacturing processes. If the past is any guide, the complexity of control being undertaken is still greatly underestimated.

The Sober Approach

A "sober approach" is not meant to imply that the modernists are drunk and disorderly, although the financial people at headquarters who pick up their capital funding requirements tend to think so. Sobriety here describes those things that can make a plant more responsive without overwhelming the problem through sheer computer power.

There are at least four ways to provide greater manufacturing responsiveness without creating a monster. They are not incompatible with a gradual preparation for modernism, and they are useful enough in themselves. They include a just-in-time method, statistical quality control, changeover teams, and design-for-production concepts.

Just-in-time Philosophy

Just-in-time, developed in Japan, works exactly counter to the traditional American approach to inventory. According to the traditional philosophy, it is useful to have process inventory throughout the production sequence in order to buffer each operation from the next. If there are enough buffers, work can continue at one location no matter what problems occur elsewhere in the system.

According to the just-in-time philosophy, it is better to remove the buffers and get the system itself working better. When the process inventory is low enough, bottlenecks can be identified and corrected. You pull out some inventory until the system squeaks; then you oil the squeak and pull out a little more. That is, you keep reducing process inventory until a production problem is identified: when that problem is resolved, inventory can be reduced further until some other stage of the process identifies itself as a bottleneck. The approach is based on a pull rather than a push philosophy.

With a push philosophy, each foreman schedules his own department's work, to optimize the mix of equipment and labor. Once a lot is ordered into production, no one is exactly sure when it will come out, let alone where it is at any one time. The traditional plant has a staff of expeditors to track down orders.

With a pull philosophy, the plant makes no product for inventory but only to customer order, and only the amount specified by the order. This places tremendous pressure for rapid changeovers on the people working the process equipment, because a batch can be as little as a single unit. In some systems a production order is released automatically by the *kanban* or work ticket that goes along with a single product kit; when the kit is assembled, the ticket circulates back to the start of the line to initiate another order.

The concept is revolutionary. It implies that the optimum-sized batch is one single unit of production. Instead of balancing setup time against inventory costs as in American traditional lot-size formulas, the attempt is to bring changeover time down close to zero and to make product as needed, one at a time. There is no risk of obsolete inventory because there is no inventory, and the plant becomes infinitely flexible over the range of product for which it has been designed.

The problem with the just-in-time method for American industry is that the underlying concept that makes the system work tends to be forgotten. "Just-in-time" is frequently interpreted by the customer to mean: "Whenever I have forgotten to order something, I can call you up in the middle of the night and you will have to deliver it at daybreak."

In reality, the underlying concept requires the customer and the supplier to negotiate very carefully thought-out requirements contracts, which will gradually bring the ordering and supply systems into conjunction. The customer defines his order pattern, lead times, and range of uncertainty; the supplier defines what lead times are possible for his manufacturing plant. The two systems are kept in close communication with each other, and there are no surprises.

If a system is based on such bedrock, it will work well for both parties. But it cannot be misinterpreted to mean "We will supply anything you want at any time, with no advance warning."

Statistical Quality Control

Vendors and production plant must be mobilized in order for just-in-time to work effectively. Delivery in the amounts specified by the requirements contract must be precisely on time, and every item must be assumed reliable because no time is allowed to test it. The number of vendors is kept to the fewest needed to ensure supply, because their output is integrated with plant assembly requirements and product specifications.

This has been a revolutionary concept for American industry, which has been encouraged to obtain bids from as many different suppliers as possible and to test everything in sight because some bad apples are always in the lot.

Statistical quality control is a method of ensuring that products meet specifications. Processes and work methods that thoroughly ensure control are developed so that no testing is required. All that is necessary are the charts that demonstrate by how wide a margin the manufacturing process can guarantee compliance with the selected performance measures.

Such an approach has become standard in the American automobile industry, where contracts include provisions that require the vendor to meet just-in-time and statistical quality control requirements. It is gradually making itself felt through the first tier of vendors to the second and third tier of materials suppliers, and across to other industries.

Interestingly enough, these new approaches have generated conflicting signals in the defense industry, where the Pentagon is obliged to see that bids are circulated as widely as possible to every potential manufacturer. This places a huge burden on the procurement system to make sure that the lowest bidders are also the most qualified.

Changeover

Changeover from one batch to the next can be accomplished by the new process technology. Changeover can also be developed as a skill by the workers, supported by manufacturing engineering teams and improved procedures. The changeover is a fit topic for quality circles; in fact, it arose along with the quality circles as a way of engaging the workers in the task of making the plant more responsive to change.

What can be done with a changeover is analogous to what happens when someone wants to change a tire on an automobile:

> When I change a tire on my car, I have to find an off-highway location where I can work; I have to look up the instruction manual to find out where they've hidden the tools; I have to jack up the car on as firm a piece of ground as I can find; I have to keep the thing level and make sure it doesn't roll downhill while I am working on it; I have to loosen the bolts from where the motor-driven machines at the factory have overtightened them; with the new tire on I have to tighten them manually to where I think they will hold; and then I have to get everything left over back into the trunk again and wipe the grease off my hands.

But things can be different:

> When I watch the Indianapolis 500 formula cars on my television set and see one of them coming in for a pit stop, they tackle this tire-changing thing differently. It happens so fast that if I blink I'll miss it. Only in slow motion can I find out what is going on—and the elapsed time includes a refueling, a windshield cleanup, and whatever else may be necessary. That's because the task of tire changing has been taken seriously and organized by a team of people who know what to do.

Changeovers are probably the least supervised task in the plant because they have been traditionally considered off-line rather than on-line work. Changeovers that were similar have taken from ten minutes to four hours, depending on how well organized the procedure had become. Yet the cost-accounting system lumped all these times together under "indirect," while every minute of the operating time was carefully tallied.

A fast changeover implies that the same equipment will be used for different products. However, another shocker for American industry has emerged in the concept of redundant facilities.

Idle machinery can be as upsetting as idle workers; but a strategy of responsiveness may well require the acquisition or use of additional machines, each one dedicated to a particular product and employed whenever necessary—with no changeover time required at all. How is the changeover time brought down to zero? By not changing the line over, but by simply turning on a different line or a different machine whenever it is needed for a particular batch.

There are occasions when it is a good deal less costly to have such "excess" equipment rather than to tear down a line and set it up again for every run, or else to carry large amounts of inventory. It is up to the manufacturing managers to recognize those occasions.

Design-for-production

The process technology is ultimately limited to what the product technology will permit it to do. If the products have been designed over time to various standards and varied components, the manufacturing plant is limited in what it can do to rationalize them. Programs to achieve product standardization are notoriously difficult to enforce once multiple designs find their way into the system. Even the standardization of nuts and bolts or electronic components can become a nightmare when two product lines in the same corporation are brought together to be made in the same plant.

It is easier to bring new products into the system under rationalized design guidelines than to try to unscramble the old ones. But many companies are still saddled with older marginal products that do not conform to any of the current design standards.

It should at least be possible to separate the new products from the old ones. The new ones should then be allowed to evolve into a statistically controlled process, leaving the old ones to dwindle and be exposed to the audits that their higher costs will invite.

DEVELOPING A STRATEGY FOR SYSTEMS CONTROL

The scale in chapter 1 helped position the plant with respect to its system control capabilities. This chapter has explored how the sales-manufacturing interface can help define the requirements and the approach. Two considerations should be kept in mind:

1. The value of alternative responsiveness capabilities to the sales function, and hence to the business
2. The options available to the manufacturing function, and the cost and difficulty of getting them working properly

Value of a Sales Strategy of Responsiveness

There is no question that economic trends are forcing increased responsiveness on all companies, no matter what the industry may be. In the defense sector, delivery requirements have always been a critical factor in contract awards; in the automobile industry they have been the basis for new just-in-time clauses in the contracts; in high technology, they have forced the manufacturing plants to become responsive to nervous and changing markets.

Since the oil shock of the 1970s there has been considerable pressure to force inventory out of the warehouses and pipelines and back into the vendor plants. As the cost of capital has increased and the risk of obsolescence has grown, it has become clear that nobody wants to hold inventory if somebody else can or if he can get the plant to make something only as and when it is needed.

The issue, then, is not whether a plant should try to become more responsive but rather how much of a task it should take on at any one time. To guide that decision, it would be helpful to document the value of alternative sales-service policies so that their value can be assessed against the costs that are unavoidably involved.

Options Available to the Manufacturing Manager

Perhaps the most significant point to make about a manufacturing strategy of responsiveness is that more options are now open to management than a decade ago. The technology is improving, both to make the manufacturing process more responsive to change and to construct a control system to master it. Essentially, three approaches are open to the manufacturing manager.

The first: make the plant equipment so responsive to change that it can deal effectively with smaller and smaller batch sizes. The success of this approach depends on the state of the process technology and how much incremental responsiveness can be brought about through equipment modernization. It is not entirely dependent on capital funding, however. Great progress in responsiveness is possible by training people to work the changeovers carefully. Some redundant assembly lines can be set up for different products so that startups can be made when the people are ready. Common and special parts can be differentiated so that scheduling can feed the common components into a number of special-assembly channels.

The second: install a fully integrated data-base materials system so that all the manufacturing functions are under complete control. Systems capabilities are continually improving. Vast computer capacities and integrated software linkages can be provided today that would have been considered impossible a decade ago. In fact, enough organizations are taking on the fully integrated data-base systems so that the task of implementing them can be organized and the difficulties determined. The difficulties are very considerable, and they are not limited to the technical side. It is not easy to get a work force that will keep to the new disciplines necessary to work and update the system—and if that cannot be done, the system will be a hindrance no matter how sophisticated it is.

The third: simplify the manufacturing task by limiting the

product line and designing for production. The success of this approach depends on how the business and its product technology are positioned. If it is possible to simplify and standardize the product, this is by far the best way to proceed. Most organizations underestimate the hidden cost of variety and most would benefit greatly from simplification. Since the approach is so valuable, one might ask why it is not more frequently adopted. The answer probably lies in the politics of industrial organizations: there are more constituencies that want variety and do not want to recognize its cost than there are constituencies that can document the cost and campaign to reduce it.

No matter how big a pitcher may be, it is always possible to pour more water into it than it can hold. But at least somebody should be minding the water tap.

6

The Engineering-
Manufacturing Interface

The interface between the product engineering and manufacturing functions becomes a point of stress whenever a new
product is introduced. With the involvement of high technology
and its continual changes in specifications and performance requirements, the interaction between the product and the process
technology can create a nightmare for the management team
trying to hold the program together. Since the business world
is becoming dominated by new-product programs associated
with some form of advanced technology, a way must be found
to resolve these problems more effectively than in the past.

Two problems underlie the introduction of new products. One
is at the management level where the commitments are made;
the other is at the operating level where the transition from development to production takes place.

THE MANAGEMENT OF TECHNOLOGY

Management faces the classic difficulty of decision making in
uncertainty. Under the pressure of competition commitments
have to be made even though no one knows realistically what
can be accomplished and what cannot. This has emerged as
the most critical issue in hyperactive industries such as the de-

fense sector, the computer industry, and any of the increasing number of industries where effective competition in international markets requires a company to offer something new before anyone else.

To begin with, product life cycles have been decreasing while the technological content of most products has increased. This has been true of virtually every field, from the obvious ones like scientific and medical instrumentation to ones such as industrial processing equipment that were once slow to respond to change.

The United States dominated the arena of high technology for some time, but it has become earmarked as a target by the Japanese. As Japan's industries matured, they shifted factory locations to the developing countries in southeast Asia, where low labor costs are still available. Meanwhile, their home industries have been modernized and they have entered the competition for high-technology products: the rate of equipment investment to gross national product in Japan has averaged 16 percent as against 10 percent in the United States.

These new pressures on American high-technology companies have placed senior management in a difficult position. They may know how difficult it is to manage the transition disciplines, but at the same time they do not want to come onto the market late with a conservative product. Their entrepreneurial juices tell them to put pressure on the organization and take some risks in order to scoop the market. So the tendency at the top of the organization is to lay down stretch targets for the organization, which are then taken as commitments whether they are realistic or not. But the people who lay down the commitments are seldom the ones who have to make good on them.

The senior manager in charge of a business that made shoes or clothing, aspirin or pistols, sewing machines or refrigerators could comprehend enough of the technology to know first hand what would be required in a new product exercise. But a senior manager in charge of a new satellite communications system or a complex weapons program cannot possibly have first-hand knowledge of the technical details. Even if he is himself an engineer, even if he pioneered the field, as a manager he is dependent on others to resolve the technical difficulties. In fact, he is dependent on several layers of management judgment, because the critical tasks may be assigned to people whose capabilities are unknown even to themselves.

So there is a vertical interaction between the senior manager eager for a preemptive success and the engineering managers directing the actual work—and it is fraught with gaming possibilities. The engineering people may play it safe and make a conservative assessment of schedules and performance in light of the uncertainties and risks. Or they may play it tough and promise stretch objectives, assuming that there will be many changes in performance requirements and budgets before they are through with the drama so they might as well get started.

Unfortunately these initial assessments tend to become wired into the planning. First, because the feasibility studies necessary to check them out are never detailed enough to confirm or deny; second, because there are so many uncertainties that it is assumed some of them will cancel each other out.

Once the program is launched, the problem of actually working the transition from engineering to production gradually emerges as the overwhelming task at the operating level. Everything in the engineering function seems to work differently than in manufacturing.

Development engineers need, or say they need, complete freedom from constraint so they can come up with creative approaches to their tasks; the manufacturing people can work only through a system of established procedures. The engineers do not know at first how they will resolve problems; the manufacturing people need to lay out every step of the way in advance. Design engineers work their way forward toward a state of increasingly clear definition; manufacturing people can work only from an already-documented base of detailed instructions.

The overall product definition is made up of a number of separate design tasks, each one of which is moving toward definition without complete knowledge of the others. Some await a delivery from outside subcontractors; others depend on a critical series of tests; still others will not be resolved until several subassemblies are put together. The product definition the chief engineer eventually achieves after such effort is only the first clarification in a series. The engineering community tends to feel that it has arrived once there is a breadboard model that works; yet it may actually require more effort to get from that stage to a worked-out design and a good bill of materials than it did to get to the working model. Much more integration and testing from many different viewpoints are needed before the product can be turned over to manufacturing and be broken up into operating procedures for production.

So the management of these transition disciplines is never a trivial task. For the general manager, the Management Row Game here takes on a dramatic and dangerous form. It is not just the trade-off between optimization of the design and efficient manufacture of the product, though that is at the heart of it. What is at stake is the management of the precarious balance between the technology of control and the technology of chaos.

THE TECHNOLOGIES OF CONTROL AND CHAOS

New product technology throws into disarray all the established systems that the manufacturing function relies on. In some areas this may increase the control available to the manufacturing manager; in others it works to throw him toward chaos.

THE TECHNOLOGY OF CONTROL

The special problems of manufacturing to support advanced technology emerged in the 1950s and 1960s with the high-performance jet aircraft programs, the Polaris submarine, and the NASA space programs. The problems continued into the 1970s and 1980s as the full impact of microelectronic technology made itself felt not only in the computer industry but also across the main front of industry in medical, scientific, telecommunications, and process controls. American industry was thus obliged to pioneer new kinds of organization and control procedures and certain new kinds of technology specifically designed to help it gain control of these tasks, which were so different from the mass-production tasks of the 1930s.

Management Organization

The functional- and product-organization structures did not provide the control necessary to manage these high-technology programs. The prime contractor had to organize teams in a variety of different technologies; he integrated the work of the major subcontractors and marshalled all the parts to make the one-off prototypes and the few-off products required by the plant; and at the same time he had to absorb a continuous stream of engineering change orders.

To deal with these new tasks a novel form of matrix management evolved. A program manager was placed in charge of a small staff and negotiated task packages through a large and varied functional organization. However, matrix management complicates organizational life. It breaks with the strong tradition of line responsibility in American industry: instead of one-person one-boss, it creates dual reporting. An engineering team reports both to its functional boss, the engineering manager, and to its business boss, the program manager. The team may be engaged in more than one program; it may run into priority conflicts between the program and the engineering managers or among the program managers.

Nobody likes a matrix structure. No one even likes to admit one has been set up. But it can be made to work. It is appropriate for certain kinds of tasks, and experienced people know how to survive with and within this kind of structure.

Much of the secret in making the system work lies in how the program manager communicates his objectives throughout his farflung and disparate empire. Every program has a different mix of priorities and set of critical performance parameters. If these critical issues can be identified in advance and realistic support for them negotiated from the engineering manager, the program manager is halfway home. The rest of the way involves keeping current with progress and difficulties without bugging the engineers. Surprises are to be avoided by both sides.

No matter how difficult the problem, if everyone involved recognizes the size of the task and finds some way to keep current with its progress, the system will work. If the engineers get bogged down and fail to acknowledge that fact, the system breaks down.

Control and Communications Systems

Two new procedures were pioneered to provide the program manager with the necessary tools: the critical path network and a cost and schedule control program.

The critical path network conceived of the entire program as a network of sequential and parallel tasks. A path through the network was based on the dates critical work packages could be completed. The procedure derived from this concept later became computerized, thus capable of being updated and linked to other tasks in the engineering organization outside the program.

The cost and schedule control system, in whatever form it took, ultimately established the traceability of all component purchases and work done so they could be assigned to a particular program. The system was built on the work breakdown structure, which identified and budgeted separately each work package in the entire program.

Both the critical path network and the cost and schedule control work-breakdowns had to be submitted at the proposal stage. Consequently both the customer and the program manager had some assurance that there was a game plan for the program.

Process Technology

As the critical path network and cost and schedule control system developed, certain kinds of advances in the product and the process technology began to make possible the manufacture of complex equipment that would have been impossible only a short time previously. Examples follow.

Sealed Components

The first computers were built around thousands of electron tubes. Since each one had a finite life, the probability of a failure somewhere in the system increased with the number of components; the first systems were off-line for repairs much of the time. Then the advent of semiconductors extended working life, lowered heat and power requirements, and the size and weight of the assembled product were reduced. Originally most of the electrical connections had to be made at the assembly plant, but with the appearance of integrated circuits they were now sealed inside the component itself. Assembling as much of a product as possible back at the component level, where it could be done under controlled conditions, became a growing trend that was applied to other industries as well.

Built-in Test

Increased complexity at both the component and final-product levels made the automation of the testing process necessary. The product had to be designed with the test sequence in mind: diagnostic circuitry was built into the product itself. An understanding of the pattern of probable failure was required for this, and as a result computer analysis of failure patterns was greatly developed so that both the product and the test equipment could then be designed to deal with them.

Computer-aided Design and Manufacture (CAD-CAM)

The age of the blueprint has come to an end, but the full implications of computer-aided design (CAD) are not yet understood. It is already clear, however, that the computer provides a simulation capability that enables the testing of a design both for producibility and stress under field-usage conditions. The design can be optimized through this process so that it can be readily assembled and tested in the factory. The computer also keeps design information in the data base. All changes are automatically incorporated in all known versions of the design, are accessible to inquiries, and are related to other parts of the design with which they interact.

Computer-aided manufacture (CAM) enables a machine to be set up quickly for a specific purpose, which is essential for high-technology programs where short runs are the rule. The mistakes possible with manual setup are eliminated, parts are pretested off-line before metal is cut, and design changes are incorporated through the programming tape.

In other words, computer-aided manufacture should become the key enabling the manufacturing function to keep in touch with the changes in product design and respond to them immediately without additional cost.

THE TECHNOLOGY OF CHAOS

Along with the technology of control has come a technology of chaos. It consists of anticipating and overcommitting each technological advance before it can be effectively absorbed into the organization.

There is much confusion between what ought to be able to be done, and what can in actuality be done. There is a lack of comprehension of the way in which each new technological toy interacts with the rest of the system. There is a gap between the potential performance of particular segments of the manufacturing process and its performance as a whole.

And above all, there is a gross underestimation of the change in the commitments imposed on the manufacturing function and of its need for stable conditions.

The Mirage of Ultimate Performance

Success in any competitive industry is based on a realistic assessment of what can be done, together with a series of suboptimizations up and down the line to achieve it.

At the very outset of a high-technology program, however, we start with a group of engineers who may not know what can be done in the manufacturing function, and may be oriented against suboptimization.

There are two unresolved problems here: first, how to communicate to the design engineers what is and is not producible in the manufacturing plant. The second, how to balance the engineering suboptimization of developing constrained designs against the manufacturing suboptimization of developing flexible resources. Neither side knows enough about the other to provide a balanced decision.

Gaming Procedures in the Determination of Requirements

All industries develop management gaming procedures when they conceptualize a new program and provide a budget for it. Yet nowhere is the art practiced with such enthusiasm, such dedication and skill, as in the defense industry.

> I was introduced to the Pentagon at a tender age, when it was still possible to make discoveries about the way the system worked. This was during the Korean War, when we were trying to speed up the procurement of materiel for the front. We pointed out a number of glitches in the system, and the people we worked with appreciated the advice and promised to do something about it.
>
> Thirty years later, the same problems are still there, though they change nomenclature on occasion. The difference is that everybody understands them these days. Every conscious person connected with the acquisition of weaponry understands how the system works and can describe the latest campaign to deal with the various difficulties. However, as one undersecretary of the Navy commented: "Getting results around here is like trying to kick around a hundred-foot sponge."

Briefly put, the system works like this: The Department of Defense wants more weaponry than it can afford. It always operates at the edge of state of the art because it does not suboptimize anything. Because of that it defines performance requirements for a new weapons system at the theoretical maximum of its potential, yet estimates the costs and time to obtain it at the theoretical minimum. It then spends its effort packaging the proposal so that the program will get through Congress.

Congress then takes the DOD proposals and chips away at the costs, eliminating the maintenance and support parts of the budget and selecting those programs with the greatest disparity between theoretical performance objectives and actual task requirements.

The ball is now passed to the defense contractors, who find themselves in the delicate position of submitting competitive proposals. This is not a situation encouraging to reality. There is only one customer, so if the contractor does not win the contract he may be out of a job. The tendency is therefore to negotiate whatever commitments seem necessary to get the business.

Once the program starts, the military generates a flood of change notices to keep the program current with the evolving state of the art. The contractors generate a few of their own to obtain supplemental funds.

Unfortunately, the military customer is represented locally by a contracting office that reports to a different sector of the Pentagon. This office was not a party to the procedures that inflated the requirements in the first place; all it knows is what it reads on the contract, and it intends to hold the contractor responsible for every paragraph.

Occasionally the contractor finds a technology that enables him to bring the program home according to specifications. But miracles are as rare as they ever were, and most programs get into exactly as much trouble as they were negotiated into in the first place.

MANAGEMENT PROBLEMS OF TRANSITIONING FROM DESIGN TO PRODUCTION

Whether the program is in the defense sector, in industry, or in a new generation of higher technology commercial products, the problems of managing the transition from design to production are severe and predictable. From the manufacturing point of view, they fall into roughly five categories:

1. Difficulties in implementing the control disciplines
2. Problems relating to program management
3. Difficulties responding to design changes on the shop floor
4. Difficulties in integrating the work of different organizations
5. Delay and disruption

Difficulties in Implementing the Control Disciplines

Critical path networks and cost and schedule control systems are appropriate tools for new-product introduction programs, but like any other tool they are only as effective as they are made to be. They impose disciplines on the engineering function that have proven difficult to accept. Design engineers are forced to define in advance how the whole complex development is going to be broken down into work packages. Each engineering manager is then able to define what is going to be done, how long it probably will take, and how much budget and manpower are needed to carry it out.

This is good discipline and appropriate to a generally defined task that needs tightening up, like an update of an existing piece of equipment. Designers not used to this discipline will go underground—they will feed into the computer whatever nominal planning data they think will keep it quiet, and then go on about their work as if no system existed. The program manager then not only has to keep current with the system but he also has to try to determine what the real status may be behind the system.

The following story illustrates the problem.

> In an avionics company, a key planning document was the drawing tree, a map of a design segment that indicates how the engineers plan to configure the design. It is not a drawing as such but rather a kind of road map showing how many drawings and what kinds will be forthcoming. Industrial engineers use these trees to estimate how many different subassemblies will be required so they can prepare the plant.
>
> The industrial engineers in this particular organization could never get drawing trees out of the design department. When the design was completed, a set of drawing trees would be issued along with the actual release of the new product for production. By that time, of course, it was no longer needed. "Once a job is finished," said one of the design engineers, "we can do a good reliable set of drawing trees for you. And don't you worry about intruding, it's no trouble at all to us."

Problems Relating to Program Management

Program management is one of the most demanding jobs in industry. It has great responsibility and little authority, since the engineers assigned to the program continue to report through their own line of command to the chief engineer. Any slip in schedule or alteration in plan during the design phase will cause endless difficulty later when the program moves into production. But the design phase is inherently difficult to manage, because the work output is not yet tangible. There are, however, signs and symptoms that are early warnings of bad things to come.

- Certain design tasks fail to meet their requirements and call on other parts of the design to yield space, increase power, or accept greater heat dissipation to cover for them.
- Work supposed to have been done in sequence, so that the later stage can build on the data base of the earlier one, is instead carried out in parallel because workers on the earlier stage are unable to complete their work in time.
- Purchasing procedures are bypassed or exceptions requested to obtain special components or subassemblies to substitute for the standard components that have failed.
- Design workarounds are made to overcome failures in specific test parameters, but there is no time to assess their interaction with other elements of the package.
- Meetings are held, to catch up on difficulties, not to plan the work forward.
- Production is started without a complete set of drawings but with the promise that some critical missing subassembly will be delivered by the time the first assembly is ready.

Difficulties Responding to Design Changes on the Shop Floor

It is important to remember that thoroughness of design and stability of requirements are particularly necessary when the plant is trying to organize its work.

Once the plant attains a work rhythm and has established methods and support systems, certain kinds of changes can be absorbed. Others are very disruptive. During the early period, before the plant is established, any change will add to the confusion; yet it is during this period that some of the most shattering changes are made.

Not all of the changes are due to immature or incomplete design. In some cases, management intervention is the cause of the problem: senior management tends to feel exposed during the interval between commitment to a new program and its crystallization into pilot production.

In industrial competition, this is the period when marketing people get nervous about any announcement of competitive products that have features not included in the company's design.

In the defense sector, this is when the military program manager realizes there are still a few combat situations the weapon was not designed to deal with. If this interval is long, as it is in the defense acquisition process, people will start tinkering with the performance requirements. Instead of reducing the design risk by allowing the product to shake down and reach some level of maturity, they want to expand its performance capabilities. Here is a typical example:

> In an aerospace company, in order to meet schedule commitments, an immature design came to the production stage before parts and subassemblies could be put through extensive testing. Established procedures were bypassed through high-level negotiation between the program manager and the senior military officer. They decided to incorporate into the weapon a new type of control unit that had performed well in a different program.
>
> When the unit was attached to the existing assembly, however, it was found to be incompatible to the design; insufficient clearance had been allowed. A larger housing was rushed into production, but it created difficulty with heat dissipation, and thermal stress caused failure in some of the electronic components.
>
> A different housing was hurriedly designed, but this time tests began to cast doubt on the structural integrity of the finished unit. A special sealant was rushed to the scene. The sealant required protection against toxicity, however, and could not be applied by the regular work force. Eventually special jigs and fixtures had to be provided and changes made in the assembly in order to use the sealant.

Imagine the cost and time implications of these changes, together with the Band-aids stuck on to repair the damage and the nurses and doctors required to change the Band-aids. If they had just left the thing alone, it would have been out in the field and working instead of inside the plant being redesigned on the shop floor.

Difficulties in Integrating the Work of Different Organizations

It is difficult enough to ensure consistent protocols between the engineering and the manufacturing functions of the same organization. The difficulties are that much greater in the large-scale programs in the defense sector, space program, and multiplant computer corporations, all of which require that work be integrated among different corporations.

Prime contractors complain they are whipsawed by their customers, but the subcontractors are whipsawed by the primes: always the last ones to be told of changes, they say, always required to meet the protocols set by the prime, and never given enough time to do their work.

The fundamental problem is the communication of design information from an organization that has evolved its own way of doing things to another that is to turn the information into hardware. Much of this information is not properly documented. An example:

> The German electronics subsidiary of a multinational corporation had developed a night-landing device for small aircraft and the unit had been put into production without any significant difficulties. A decision was then made to transfer the product to the English subsidiary of the same corporation. A complete set of blueprints and all the necessary jigs and fixtures were transferred to the English plant, together with a manufacturing engineer assigned to help them prepare the facility.
>
> The moment the German engineer left, the English plant began experiencing difficulties. Component parts, which were apparently working well enough when tested individually, failed on assembly; operating sheets did not seem to match the actual requirements; dimensions on one blueprint did not conform to the dimensions on another. English components, made to the same specifications as the German components, tested differently or had different side effects when matched with the rest of the unit.
>
> Gradually it became apparent that the documentation had not been kept up to date. It was in a variety of stages of disarray, and considerable "lore" was involved in getting the unit assembled properly and working.
>
> Eventually the English engineers were obliged to redesign the unit and develop their own operations sheets and components lists. They got the product working, but it was a different product than the "Chinese copy" they had hoped to make.

Delay and Disruption

A basic rule to keep in mind is that if anything is likely to go wrong, the factory is the most costly place to put it right. Early on and upstream in the flow is the place to check things out.

The same correction will cost many times over if it comes late in the sequence, when the product is being assembled. A case in point from defense:

> A minor change was being made to the electrical system of a submarine under construction. The amount of havoc this little alteration caused was unbelievable: in order to get at the electrical unit, it turned out that they had to rip out some of the pipes that were covering it; in order to get back to where they had started, they had to retest the entire hydraulic system and replace a good deal of the wiring. By the time the new electrical component was installed, the whole submarine schedule had been delayed.

The same thing can happen in industry:

> I once saw luxury automobiles being manufactured. I couldn't understand why they didn't drive the things off to the distributor when they came off the assembly line. Instead, they kept those cars around for weeks, testing and re-testing them. At first I was terribly impressed. Then I began wondering why, with all that value added to the product, they had to futz around for weeks after the automobile had already been made.
>
> I made a study of the test procedures and discovered that the actual testing time was trivial. What was going on was postassembly manufacture, delay, and disruption.
>
> A car would be held until some part that had failed to arrive on time was retrofitted to it. But by that time the technicians had to disassemble some of the automobile to fit the part in properly. In doing so, they would detect a little looseness here or a lack of finish there, so they would open up a whole section of the car to remanufacture it.
>
> I was able to observe a lovely little ballet with one of the doors. There was a harshness in the window crank assembly. So the technician separated the inner door from the outer door to tighten the mechanism. While doing that he loosened some of the wiring. But that did not show up until a later test. At which time they opened up the door again.

TAKING CORRECTIVE ACTION

The anecdotes illustrate what happens when human nature is left to cope with a vast production system that has broken down or been thrown off balance before it can get to its proper working momentum, or else is working fine until somebody steps in and tries to touch it up a little.

Is there a pattern to these difficulties? Can management anticipate the problems and set up measures to deal with them? The answer is yes, there is every opportunity to take corrective action.

The bulk of the difficulties in the transitioning process from engineering to production are identifiable in advance; by this time enough has been learned to organize the procedures to constrain them.

Identifying the Transitioning Difficulties

In the early 1980s the Defense Science Board, an advisory group to the Pentagon, organized a task force to examine transitioning difficulties in the defense industry. The task force was made up of experienced people from both the military and industry who believed they knew what the transitioning difficulties were. They decided that industry did not need one more study to find out what was already known but rather a reference manual to checklist the problems and suggest corrective actions.

Their report was published in 1983. It has become a classic statement of the problems and has led to subsequent versions that extend the coverage and add to the corrective "templates" for the problems. In 1984 an industrywide conference was held on producibility, sponsored by the Manufacturing Management Committee of the National Security Industrial Association, supported by the Research and Engineering Committee. This group of industry leaders adopted the general thrust of the task force and reported on the progress of the transitioning task within their own companies. In 1985 the Department of Defense began to use the report's criteria as checklists in the award of contracts.

The task force identified 29 areas of risk in the transition from development to production:

1. Funding early in the program inadequate to permit effective engineering, test, and production planning
2. Intangible design requirements that are not directly measurable during the design process

3. Lack of understanding of the mission profile—that is, the complete set of functions the product will have to perform and the conditions under which it will have to do so
4. Little use of trade-offs to determine the relative value of certain features against the producibility and reliability of more simple designs
5. Lack of corporate design policies, including specifics for designing to reduce risk
6. Parts and materials utilized at their maximum theoretical stress levels, rather than allowing a safety margin
7. Specialized design-risk analyses not made by the design engineers most familiar with the product
8. Design reviews that lack specific direction and discipline and become one more nominal checkpoint to pass
9. Insufficient use of built-in test circuitry
10. Designs that require skilled model-shop technicians to assemble and do not survive rate production in the factory without degradation
11. Designs that do not allow for access by inspectors and automated test equipment
12. Designs developed without knowledge of manufacturing processes, vendor parts availability, or length of run required
13. Unbalanced test programs that call for excessive testing of certain functions and inadequate testing of others
14. Tests that do not consistently reflect field conditions
15. Reliability development testing not tailored to the needs of the particular program
16. Inadequate use made of failure reports to influence design improvements
17. Inadequate feedback from field use
18. Manufacturing plan not developed early in the process in conjunction with the design engineers
19. Inadequate receiving inspection made on critical component parts
20. Inadequate involvement of key subcontractors in the design, development, and production planning process
21. Inadequate systems to track information about defects
22. Inadequate environmental stress screening to identify patterns of poor workmanship

23. No common data base between design and manufacturing to integrate instructions from computer-aided design and computer-aided manufacture
24. Interruptions in the schedule of production, leading to a loss of learning by the factory
25. Lack of a specific and detailed transition plan
26. Lack of modern facilities and equipment
27. Excessive dependence on manual assembly
28. Unstable work force
29. Inadequate personnel planning within the program office

The list has been fine-tuned a little since it was issued, but it has never been seriously questioned as identifying the critical difficulties in the transition from design to manufacture. Discussion has always centered on the question of what to do about it.

Taking Management Action

Four kinds of management actions are possible.

1. A technological strategy can be developed that will place bounds around the program mission.
2. A risk assessment can be made to anticipate the critical factors in the program and prepare the means to contain them.
3. Transition disciplines to manage the interface between engineering and manufacturing can be enforced.
4. Product and process technology can be combined to lower the risks of transition.

Technological Strategy

A business organization can be structured to do almost anything, provided it is adequately prepared. In some industries survival requires high-risk pioneering; in others it is possible to position the company so that products brought out are improved versions of tested designs. Either strategy can be implemented, but the organization has to understand what it is up against; the change from a conservative to a high-risk strategy is not likely to be uneventful.

For example, in the American automobile industry, the model year traditionally sets the pattern for changes. Cars are cosmetically changed every year, with a heavy redesign every third year or so, and all changes are keyed to the introduction of the new models in the September shows. In the European car industry, however, changes are not traditionally keyed to the model years. Some companies continue the same basic model for a decade or more, incorporating changes as they desire them to be introduced at any time selected by the company.

> After a long period of stable production with little design change, a completely new model was introduced in 1960 by a European automobile firm. The disruption nearly bankrupted the company, because the existing system was not capable of responding to what had become in American companies a fairly routine task. When the same company planned a new model in 1975, it spent five years preparing for the change and managed it successfully.

A technological strategy clarifies to senior management, and to the engineering and manufacturing functions, the nature of the company's objectives. It identifies those elements that look new but that are only extrapolations of existing technology; it identifies the pioneering areas and assesses what kinds of skill or knowledge are necessary for success.

Such a strategy is not limited to the product technology but rather considers the product and process requirements together, as a single entity. It identifies where in the corporation the critical skills reside, which prevents the company from losing its corporate memory and putting that part of the organization onto some other task while hiring new people for a program that requires those same skills.

The strategy assesses alternative approaches that stretch the technology farther or that relax some of the objectives in favor of a more gradual approach. It makes a preliminary assessment of time, people, resources, and new funding needed to carry out the strategy. In part this is insurance against drift into deeper technological waters once a program has been launched.

In other words, the technological strategy lays out in a comprehensive way what it is that the organization is trying to do. A risk assessment can then be made and risk management practices can be prepared.

Risk Assessment

With a clear definition of the nature of the technological reach that the company is trying for, most organizations can anticipate the problems and set up methods of containing them. If the key people cannot do that, it is a sign they do not understand the nature of the task they are taking on.

The objective of the manufacturing function is to gain access to the decision-making and design-formulation processes before commitments have crystallized. This entails three difficulties.

First, the manufacturing manager is not always brought into senior management discussions about new program activities until the program has been conceptualized and clearance is wanted. That is, senior management as a rule wants to hear that the manager can make the product with a little bit of this and a little bit of that by way of additional resources; they do not want intervention or the suggestion of a different version.

Second, penetration during the early stages of design is difficult. The design engineers typically say that it is too soon to discuss anything meaningful because they are still dealing with many different possibilities. Later, they will suddenly produce a design and try to protect it.

Finally, in a lot of cases the manufacturing managers do not have much to contribute. They can react to product designs once they have something to examine; but they are not pro-active enough to provide guidelines for the designers.

If risk management is to be truly effective, the designers must be able to identify areas of concern in advance; they must know in what ways they are trying to push the existing components or performance levels beyond the normal stage of improvement. The manufacturing people must be able to identify in advance the kinds of trouble they are likely to have with a new stage of high-performance product. Where will it occur?

- The component parts: Will the latest electronic components be able to perform the way the advance specifications say they will, or will the new components have to be tested one at a time to make sure they meet the specs? Will they be obtainable on the dates required, or are they going to hold up the whole program?
- The subcontracted sections: Will the product performance depend on certain critical units being made effectively by outside organizations? If so, does the company have any expertise in assessing those organizations' competence or confirming their progress, or will it be necessary to wait until they are good and ready to deliver a "black box"?
- The structure of the assembly: Can the components be assembled with the regular work force using existing tools and fixtures? Or will they require some kind of special sealant, compression fit, or heat-dissipation system before they can be made to work?
- The test program: Does the company know how to test for the performance factors aimed at? Are there interactive elements that will be apparent only on final assembly? Will the specifications for the subassemblies and components ensure that the final product will work? Will the company be able to live with these specifications, or are they going to require entirely new equipment and methods of production?

Risk management is an art not yet practiced with skill or assurance. But these can be developed, particularly if each program is not looked at as if it were entirely different from anything the company has ever done before. The questions identified above may be answered in a lessons-learned study of previous programs. They can be focused by establishing lists of suppliers and types of components that have—and have not—been reliable in the recent past; by identifying specifications and tolerances that the manufacturing process can meet as is and those that require special equipment.

Enforcing the Transition Disciplines

A good deal of the outcome of the program depends on whether the transition disciplines are enforced. This requires a program manager who believes in enforcing them and an engineering manager who takes the issue seriously and feels some responsibility for the final product. Above all, it requires a proactive manufacturing manager.

By proactive is meant a manufacturing manager who guides the engineers rather than interferes with them. It does not mean a manufacturing manager who can point out what has gone wrong once the design is already in production.

> An aerospace plant was shifting from an analog to a digital computer system for a satellite assembly. The manufacturing manager examined the prototype and found so many soldering connections and so much harness work that he knew it would be both labor-intensive and confusing for the workers. He foresaw a squad of inspectors trying to get a battalion of assembly workers to go back and solder the correct connections.
>
> He held a series of meetings with the chief engineer to discuss the problem. They finally devised a producibility rule of their own: "Thou shalt not design a circuit board with more than ten wires issuing therefrom!"

The Technology of Control

Management can seek out and develop the technology of control and try to avoid the chaos. But control has to be considered a common task, combining product and process. It does little good to be at the state of the art in product design if the product has to be fitted together with hammers.

There are some unresolved tasks here. Computer-aided systems are not yet integrated into the design and manufacture process. The design community is not yet very knowledgeable about production processes, and the manufacturing community is unable to provide effective guidance in the early stages of design. The software guiding the designers still does not talk to the software guiding the machines. Ultimately this is a management problem, because its solution depends on design engineers and manufacturing people who know how to integrate their activities with each other.

Moreover, product design does not make full use of sealed components, standard packages, and modular parts. Product simplification, so that fewer parts are used, has a far-reaching effect much beyond the obvious one. It limits the number of items that have to be defined, purchased, inspected, and inventoried; it limits the number of interactions among them; it limits the exposure to error and the buildup of indirect costs all the way through the line and for the life of the product. Yet almost any product subjected to value-analysis reveals that it could be simplified from its present design.

Finally, process technology relies too heavily on manual tasks. Manual operations are considered costly because they require direct labor input. But the real costs build up when the operations are subject to error; then the hidden factory of supervisors and inspectors begins to take over. The new generation of flexible machine systems should be capable of doing many of these jobs correctly the first time. Unfortunately much industry is still laggard in the funding of plant modernization.

Management of the technology of control requires three things.

First, design engineers must understand how their work will take shape in the factory. They have to feel responsible for seeing the product all the way through to production. They have to anticipate manufacture, be open to discussion with manufacturing people early in the cycle, and organize an orderly transition to production once their designs begin to take shape.

Second, manufacturing people must learn how to propose specific guidelines to the engineers without interference. They have to learn to define what it is they can and cannot do with their existing resources. They must understand alternative designs enough to suggest ways of optimizing reliability and cost-effectiveness through their own plants. They have to be willing to involve themselves in design discussions early in the cycle.

Last, senior management must face enough operating realities to know when they are listening to honest advice. They have to know what to do when someone says that a timetable cannot be met or a design is at risk. They must learn how to open themselves to advice—and how to assess that advice so that they make strong decisions and take risks without marching blindly into a disaster.

Part III

Integrating Manufacturing into the Business Strategy

7

The Workshop Process

The Management Row Game offered a conceptual framework to test the feasibility of the business strategy in industrial corporations. It did so by providing a means of:

- Examining the business strategy from the point of view of the functional resources necessary to support it
- Identifying the trade-offs between functions by referring difficulties back to the business strategy as a means of guiding priorities
- Assessing progress in the functional organizations by relating it to their ability to support the longer-term business objectives
- Assessing the viability of a business strategy in competition by determining how flexible the functions will be in the face of sudden contingencies.

On the whole, the Management Row Game provided a conceptual framework but no working process by which it could readily be applied. What was needed was a kind of shakedown procedure through which a profit center within a corporation could determine what resources it needed in order to carry out its commitments and make them known to key personnel in the functional organizations.

It was at this stage that the group manager of a major industrial corporation encouraged us to get to work in some of the businesses reporting to him, most of them in the defense and automotive support industries. He was barraged by demands for capital funding for plant modernization but he was not convinced that the manufacturing people would spend the money to do more than update existing equipment. He wanted to find out how their plans could be dovetailed with the businesses—what alternative strategies could be carried out with greater or lesser funding and what it was they required to "get their act together" in order to achieve stretch objectives.

It was in this context that the workshop process was developed. It has since been applied in other corporations to approximately 20 different businesses.

THE CONCEPT

We wanted to bring together all the managers connected with a strategic business unit for a two-day period during which they could work in an unthreatening atmosphere. From 40 to 70 people would be involved, so we had to make sure that there was some mechanism for bringing up significant issues and opportunity for a general discussion of them. At the same time we tried to create something different and less threatening than the planning review process.

Planning reviews are formal occasions in most corporations: a small number of business managers present their plans to a senior management committee that sits in judgment; the managers try to win approval for their plan and obtain capital backing for it. The atmosphere is inherently confrontational and competitive and not everybody comes out feeling like a winner.

The workshops we organized were to be used solely for the benefit of the business managers. Senior people might be invited to attend, but with the understanding that they should be well-behaved and unintrusive.

Furthermore, it was to be clearly understood that no decisions were to be made during the workshops. They were discussion forums to focus issues and calibrate priorities—an open meeting conducive to active debate. Although we did not want to confront the general manager with commitments then and there, we did want the workshop to lead to clear action within two or three weeks; at the least we wanted to see task forces organized to resolve some of the more critical issues that had been raised.

From the beginning our most important objective was to get all the managers together in one room so they could speak freely about their own concerns for the business, understand where the other managers were coming from, and discuss alternative approaches openly.

In order to provide this relaxed atmosphere during the meeting, however, a lot of effort had to be applied beforehand to prepare every step of the way.

THE PROCESS

The workshop process is a disciplined procedure designed to help middle managers in the different functions work together effectively to help implement the business strategy. It clarifies the principal capabilities and constraints of the functional organizations and the nature of the support that each one needs from the others in order to get its own work done. It opens up discussion of what is needed from the functions to implement the business strategy, what conflicts might develop for the use of scarce resources, and what realignment of resources—in capital, staffing, time, or knowhow—might be desirable to better implement strategy and avoid conflict with competing requirements within the corporation. Finally, it leads to action that may reshape the functions to support the business strategy and that leads to the dedicated support necessary to carry out a strategy within a corporation where there are other commitments competing for scarce resources.

The workshop process consists of six stages:

1. The business unit is defined, the key players are identified, and the business strategy and expectations are determined from senior management.
2. Field interviews are conducted with the key players to determine the critical issues that affect their ability to carry out the business strategy.

3. Discussions with senior managers and the general business manager lead to a plan with agreed-upon principal issues that focuses on key areas of concern.
4. A limited number of key functional managers are selected to make presentations during a two-day workshop that is to be attended by the managers and staff connected with the business.
5. The workshop is held. Both the presenters and the nonpresenters participate actively; issues involving the implementation of the business strategy are discussed and debated and problem areas are identified.
6. Interfunctional task forces are then organized to take action on a limited number of the most critical problem areas.

Defining the Business Unit

In order to apply the process, it is first necessary to carve out the territory—and it was surprising to find in how many cases it was necessary to clarify what it was that constituted the business unit.

Since many businesses in conglomerate corporations are obliged to share resources, it is often necessary to redefine what constitutes a separable business unit for management purposes. There are business groupings made for the purpose of simplifying financial control at the top of the organization that do not in fact constitute homogeneous management clusters. There are businesses that contain quite separate market segments that are linked by a common technology. Others comprise related markets but are supported by unrelated manufacturing operations. Four criteria were applied to identify separable business units: critical mass, technology, special character, and the need to clarify the boundaries.

Critical mass: Is the business large enough to be treated as a separable unit for management purposes? Does it have the longevity and market potential? Or is it perhaps so submerged within a larger organization that it is not getting adequate management attention?

Technology: Does the business require a distinctive technology, either in product development or manufacturing? Has the business been able to attract and retain an adequate staff of dedicated technical people? Are they enough in contact with each other to provide the kind of combined weight that will lead to a professional discipline within the larger organization?

Special character: Is the business obliged to move against the mainstream to obtain the kind of support it needs? Is it, for example, a specialty business within a mass-production organization? Or is it a mature cost-constrained business operating within an organization dedicated to some new high-growth technology? Is there some special flair or market knowledge that this business must maintain that is in danger of becoming lost in the general scramble toward other objectives?

Need to clarify the boundaries: In virtually every workshop, the initial list of interviewees omitted critical people. This is not because the senior managers did not know their organizations but because a number of important links in the chain are not always obvious in the power structure. Only when the product flow is methodically traced through the system, from the purchase of raw materials to the sales and service of finished product, do the gaps show up. An example:

> The inventory in this business was always out of balance because the plant was never able to meet delivery dates. The production people would stop what they were doing and make special runs to satisfy certain customers; by the time they got back to their regular production they were late with that, too. Everyone assumed that the delivery dates on the orders were set by the customers and that late was late. In the process of interviewing for the workshop, however, we discovered that the customers did not always set the dates, nor did they specify whether the order was to keep a plant in operation or just to replenish stock. We finally traced the problem to a young fellow in a back room of the home office; he was putting delivery dates on all the orders that lacked them. When we asked him on what basis he was specifying the dates, he said that he put down whatever he thought looked reasonable and that it must be all right, because nobody had ever told him to do anything different!

Sometimes it is necessary to "create" a business segment, or at least to redefine it:

> In one worldwide corporation, business planning was dominated by the American marketing people. They defined objectives, product require-

ments, and by implication the manufacturing resources needed to back up the product line, all on the basis of the domestic requirements for which they were responsible. It was found, however, that the foreign subsidiaries made demands that accounted for a substantial portion of the production output. This meant that the plant was being provisioned on the basis of a domestic demand that was only two-thirds of its actual worldwide requirements.

Another example:

An organization made a line of products then sold by a second organization within the same parent company; the second organization made a different line of products, and that was sold by the first. Because compensation plans and priorities differed in each, the situation led to a distorted emphasis by both sales groups. One tended to pick out the commodity products and sell them on a price basis; the other tended to pick out the specialty products and sell them with technical support. They found that salespeople from the two organizations had been transmitting some very mixed signals to the customers.

Conducting the Field Interviews

The main purpose of the field interviews was to determine how well the functional managers understood the difference between the long-term objectives of their own organizations and the requirements of the different businesses for their services. Our discussions with them move along the following lines.

Work Objectives

We start with the interviewee's place in the organization. We want to see how he fits into the organization—whom he reports to and whom he works with, and what kind of resources he is able to deploy. We want to know what he is trying to accomplish, what obstacles lie in his way, and what he thinks he needs to help him accomplish his objectives.

Business Strategy

We then try to find out how much he knows about the business strategy and how the functional work relates to it. And we have had some surprising responses: people who do not know or who have misinterpreted the business strategy; people who acknowledge that their own functional objectives are moving in a different direction from those of the business; managers who seem surprised that there should be a connection between the capability of their functional activity and the performance levels that the business manager needs. And we find some managers whose work supports so many different business units that they have virtually no identification with any strategy at all.

Functional Interrelationships

We want to know which other functions the interviewee has the most to do with and what kinds of issues come up in working with them. How are these problems resolved? How effective are communications with those functions? What kinds of things does he need to know to do his job effectively, and what kind of advance notice does he need to utilize the information effectively?

We generally get good responses from these questions. However, when we ask what kinds of information the other functions need to get from the interviewee to carry on their work, the answers tend to become vague.

Dealing with Change and Uncertainty

We want to know what kinds of uncertainties the interviewee is obliged to work with and how the changes arise, whether as a result of external conditions imposed by customers, suppliers, or competitors, or from changes in internal management policies or priorities. We want to know what kinds of changes were caused by the other functions and how he was informed of them.

How does he deal with these changes? Which ones are the most difficult? What time factors are involved in his ability to respond? In what ways does the organization or the reporting system help or hinder him in responding within the required timetable? What is the effect of his work on the other functions?

Setting Priorities

Basically we want to know how he uses his time. What are the principal routine activities? How much time is spent "fighting fires"?

Setting Performance Expectations

How does he define the performance criteria of his function? How does he communicate these standards to the people who report to him? How does his boss let him know what is expected? What performance is it reasonable to expect if his function supports the strategic objectives of the business as fully as possible? Is his function at the state of the art, considering best professional practice on a large scale?

This last question elicits particularly interesting responses. It is disheartening to learn how little some manufacturing managers know of best practice worldwide. They may not know what their competition is doing, what other plants in the world are worth visiting, or what professional meetings they might attend to learn more about the state of the art.

Assessing Resources

If he had more resources in equipment and staff, what would he be able to do with them? This question must be prefaced by the observation that he is not about to get what he wants—at least not through the workshop.

If his budget were cut back, what minimum kinds of service and support would he be able to provide? This puts the issue of resources on a more neutral basis than the previous question, and the interviewee can sometimes give a good explanation of flexible budgeting in his department.

Planning the Workshop Agenda

A workshop can be used for a number of objectives. One possibility is to keep the workshop focused on implementation of the business plans as they have already been laid down. Another is to lay down a general strategic direction and ask for the preparation of functional action plans to implement it. Still another possibility is to pose two or three strategic alternatives—usually depending on the "reach" or "risk" that the managers are willing to support—and assess problems and opportunities. And another is to open the strategy discussion to any valid proposal and concentrate on getting the interfunctional relationships to work better.

In order to get our signals straight and determine what to aim for, at this point we reviewed the situation with senior management. We wanted to know what issues might be ruled off limits for discussion by middle management. These may include the financial criteria for business performance. It may also include strategic alternatives not within the range of middle management to influence, such as acquisitions or spinoffs, organizational structure, and integration or separation of the business with (or from) other divisions of the corporation.

Priorities

Any business will have a hundred things to discuss, but a two-day session cannot cover a hundred topics. For that reason a good deal of thought is invested in keeping the agenda focused on the half-dozen or so issues that can really make a difference to the business.

These priority issues will have been discussed during the course of the interviews, and everyone will mistakenly suppose that everyone else interprets the issues in the same way. The agenda provides an opportunity to redefine them and establish some kind of common language for the workshop. It is important to recognize that too much on the agenda will reduce the impact of the critical issues. In this process, less comes from trying to do too much.

Most of the workshops have had one dominant theme. Here are some examples of kinds of businesses and their themes.

- A business heavily committed to mature markets but with new products in each area: the theme is the reallocation of resources to maintain existing positions with minimum resources while shifting support to the growth areas.
- A business that has been part of a larger and more demanding business segment: the theme is to examine it as a separate activity and determine how best to exploit its potential.
- A business with seven plants in North America, each one in a different stage of progress or disarray: the theme is to develop

the criteria for rationalizing the disparate product line and reducing the number of plants to four.

- A worldwide business with very difficult problems of new-product introduction: the theme is to develop a model for working the problem across national boundaries by integrating the marketing-engineering-manufacturing capabilities more effectively.
- Two business lines with a market and a plant that cross both: the theme is to clarify responsibilities and needs so that both markets can be effectively supported.

Launching the Workshop Presentations

Once the signals are straight, a workshop agenda can be built around a limited number of presentations by some of the key functional managers involved with the business, and the workshop itself is launched. This involves three activities: preparation of the agenda, the kickoff meeting, and dry runs.

Preparation of the Agenda

A business can be looked at "from the outside in" or "from the inside out." That is how we distinguished the two days of the workshop.

On the first day, we want to understand the outside world to which the business strategy is responding. This means a summary of the strategy, an assessment of the market, an evaluation of competitors, and a financial review of the business. It means coverage of both the marketing and the sales activities by major product segments and a summary of the new technology being developed to bring new products into play during the subsequent three to five years. We need to know what commitments are being placed on the organization.

On the second day, we want to look at the business from the inside out in order to understand how the business is going to meet the commitments imposed by the "outsiders." This means an understanding of the process technology and its ability to keep up with competition and the state of the art. This means the manufacturing plants need to be understood in terms of their capacity, ability to respond to change, ability to meet quality and performance commitments, and the configuration of their equipment compared to the width and range of the product line. It may examine the product mix in a typical schedule and assess the ability to manage inventories within the kind of uncertainty typical of market conditions.

Figure 7-1 is a list of presenters taken from two workshops, one in a process industry and the other in a company making high-technology industrial products.

The Kickoff Meeting

The presenters are invited to a half-day kickoff meeting held five or six weeks before the date of the workshop. The conceptual framework is explained, the objective of the workshop is defined, the group vice president and the general manager of the business

7-1. List of presenters for two workshops.

Process industry

First Day: Outside In
1. Business General Manager
2. Controller
3. Marketing Manager, Commodity
4. Marketing Manager, Specialty
5. Central District Sales Manager
6. Salesman, Specialty Products
7. Specialty Research Program Manager
8. Sales Service Manager

Second Day: Inside Out
1. Operations Manager
2. Plant Manager, New Jersey
3. Area Manager, New Jersey
4. Shop Superintendent
5. Inventory Control
6. Quality Assurance
7. Industrial Engineering
8. Scale-Up
9. Purchasing Manager
10. Plant Manager, California

(continued)

7-1 (*continued*).

Industrial products

First Day: Outside In
1. Group Director
2. Group Marketing Manager
3. Group Financial Manager
4. Program Manager
5. Program, Development
6. Program, Test
7. Program, Quality Assurance
8. Chief Engineer
9. Technical Group A
10. Technical Group B
11. Technical Group C

Second Day: Inside Out
1. Manufacturing Engineering
2. Subcontracting, Group B
3. Manufacturing Management
4. Scheduling
5. Materials Controls
6. Electronic Assembly
7. Machined Parts Assembly
8. Test Manager
9. Field Support

go over their expectations, and each one of the presenters is given an outline of an agenda that indicates the kind of coverage he is expected to provide and the time limits. The last are always severe because of the need to leave ample blocks of time for discussion.

The rest of the half day is spent in discussing the presentation agendas so that the group will see how each one of them fits into the program. The emphasis is on the most critical issues and on each person's ability to communicate to others not familiar with his function. The main rule is to explain one's own problems in someone else's language so that he can understand how the work relates to him.

Figure 7-2 is the agenda for the manufacturing presentation. This is backed up by a hard copy of the material distributed to

7-2. A manufacturing presentation (80 minutes).

1. General plant array (10 min)
   Plant locations
   Incoming materials flows
   Outgoing product flows
   Product specialization by plant

2. New Jersey plant (5 min)
   Overall plant capabilities and constraints
   Businesses served: combinations and separations; support and interferences
   Manufacturing mission for specialty products group B

3. Process flows for specialty products group B (15 min)
   Incoming materials
   Kitting
   Machined products
   Electronic products
   Subassembly tests
   Final assembly
   Shipping

4. Cost–volume relationships (10 min)
   Cost structure in the recent past
   Analysis of overheads
   Effect of 30% volume increase
   Costs at optimum operating conditions: in both volume and product
   mix

5. Product line changeovers (5 min)
   Present batch run configuration
   Number and cost of changeovers
   Preferred sequence of changeovers
   Optimum product line campaign

6. Quality strategy (10 min)
   Recent performance: right the first time, rework, and scrap
   Critical problem areas
   Objectives
   Statistical quality control program status
   Long-term quality strategy: specifications, product, process, control system

(*continued*)

7-2 (*continued*).

7. Problem products (5 min)
    Half dozen most difficult products
    Problem areas
    Help needed from other functions to resolve them

8. Response capability (10 min)
    Recent performance
    Definition of the most responsive configuration
    Cost and effort required to reach the most responsive objective
    Problem areas
    Cost savings obtainable from relaxing responsiveness commitments

9. New product introduction (10 min)
    Incidence of new product introduction in the recent past
    Assessment of cost and effort requirements
    Problem areas
    Support required from engineering to resolve some of the problems

the entire group after the presentation. By doing this the workshop binder can become a reference document.

The Dry Runs

After the kickoff meeting is concluded, the critical period is the last three weeks prior to the workshop. Phone calls will be made during this time to make sure that people are preparing, and a round of interviews is scheduled to go through dry runs of the presentations.

Experience indicates that this round of dry runs is important in helping the managers prepare for the workshop, in utilizing hard evidence, and in keeping them focused on the most critical issues. Unless they have this chance to talk through their presentations, the results are likely to be a very mixed bag.

Conducting the Workshop

The key to the workshop process is the fact that it is structured to cover as much ground as possible, and to invite participation from as many people as possible, within a two-day period. Despite the structure, there have been surprises.

- The shift foreman who told the senior people what it was like to keep the obsolete plant facilities tuned up and operating with bailing wire and adhesive tape
- The field salesperson who said he didn't know why any customer would want to buy one of the products because it had been outclassed by every other competitive product made here or abroad
- The creator of a brilliant marketing strategy who caved in on cross-examination and revealed that there was, in fact, no data to back up any of his assumptions
- The quiet research manager who pointed out that his development program was ahead of schedule and that the new technology he was working on would be more of a breakthrough than anybody had hoped
- The sales administrator who covered the steps necessary for a customer to receive adequate service—and showed the group that the much-vaunted technical support service had become bogged down by bureaucracy
- The industrial engineer who walked the group through the process of transition from design to production and identified, one after another, the points to where the product reliability and delivery problems could be traced

The workshops have tended to be more different from one another than originally anticipated. Even within a single industry, the businesses in the corporation were very different from each other. Once the agenda was adjusted to individual needs, the workshops evolved along their own natural lines.

In one case, a supportable strategy had been laid down and the agenda was devised to help the functional managers work out the action plans; in another, the strategy was picked to pieces by one serious functional limitation after another.

In another case, one of the major issues driving the workshop was resolved easily and another emerged as the key problem. Prior to the workshop, most of the people we talked to had been concerned about the capability of one of the major process units to meet competitive quality levels. Everybody had been agonizing about it and predicting dire results if the problems were not solved, while the technical people were arguing that they could not be solved on the existing equipment. But when we opened up the issue at the workshop, we got one favorable report after another—a new machine drive was being put in place, the instrumentation on the equipment was being improved, the most recent batches were fine. However, the pricing of the product had deteriorated under an avalanche of cheap foreign imports and the whole issue surfaced again under a different guise— how to speed up the throughput of the process unit so as to reduce costs.

We always made the point that the workshop was not a decision-making forum. The objective was to focus knowledgeable discussion on the key issues, not to try to resolve them during the meeting. However, as they evolved the workshops prepared the path for decision-making in a more direct way than we had anticipated.

It is the responsibility of the facilitator to summarize carefully and in detail. We formed the practice of summarizing at the end of every discussion period so that there would be less pressure on the final conclusions, where the facilitator might unwittingly impose his own views. However, the summaries tended to have a cumulative effect, so that by the end of the meeting there was generally a sense of direction. Either it became clear that some issues were resolvable and others required a great deal of combined effort, or it was clear that some strategic approaches were based on fishy assumptions. Perhaps the business had a winner of some kind that needed to be vigorously exploited. Perhaps the current program was perfectly sound and it was clear that everybody should support it. In some cases at least, the workshop process itself led to a sense of conviction as to how to proceed, after which the decision making was more routine.

In addition to the general manager of the business and his team, the group vice president often came to the workshops, sometimes with teams of senior corporate officers who stayed for short periods of time. The group vice president generally opened the proceedings and had the last word in closing them.

Occasionally he would participate enough to give an opinion on corporate policy or else to warn people off certain approaches that involved considerations of turf outside his domain.

The Task Forces

After the workshop ended, the business managers got together to assess the state of play and decide on the best means of taking action on the unresolved problems. This is the most important part of the process, and the one that we have had the most difficulty with.

Some organizations already had in place the structure and communications systems necessary to deal with the kinds of issues that come up. Others had to establish special task forces.

The workshops tended to highlight issues that apparently resided with one or another of the functional activities but that could not be resolved without the active participation of the others. In fact, when the task was defined clearly enough to make it capable of resolution, it usually became clear that there was a multifunctional problem that went beyond the regular scope of functionally organized activities.

When a task force was established, it required very careful preparation and guidance if it was to come up with solid results. Whenever the task forces were just thrown together, we found that they wasted much of their time at least during the early weeks of the meetings as they tried to get organized.

We evolved some general rules about the task force organization and procedure:

- It is important to provide a task force with a clear charter that lets them know at the outset what is expected of them. Are they to propose alternatives? Are they to recommend organizational changes or keep away from them? Are there any corporate or business policies that they must keep within?
- They must come up with a timetable that meets certain expectations within a limited time frame, usually three months. This means that they cannot go into exhaustive detail even on important issues, and they cannot allow themselves the luxury of following up a lot of related side issues.

- Whatever they do come up with, it has to have an action orientation. This means that the power bases needed to carry out the action probably should be represented on the task force.
- The chairman has to be someone who knows how to use agenda, milestones, and interim assignments to get results out of a group of people who have other things to do.
- There should be a limited number of task forces (four or five) and a limited number of people in each group (four or five). They can co-opt other people temporarily to provide themselves with supporting data if they have to.
- The task forces should be interlocked by a steering group consisting of the general business manager and the chairmen of the task groups so that they do not make recommendations that are mutually incompatible. The steering group has to make sure that they are more or less on schedule, and that their recommendations will lie within some band of practical action that the company can endorse.

8

Results of the Workshop Process

The workshop process, as a disciplined technique to implement row game management in industrial corporations, has been applied to about 20 different businesses. What emerged from the experience can be summarized along two lines: first, the kind of companies to which the approach was found applicable; second, the kinds of issues that were raised and the actions that were taken.

WHERE THE MANAGEMENT ROW GAME APPLIES

The Row Game is useful in any situation where the functional organizations dominate the business organizations that cut across them. The workshop process is helpful in integrating functions to support a business strategy in situations where different businesses must share common resources.

A number of patterns have been identified:

- Companies essentially organized along functional lines but needing a more rapid response time to meet competition or to exploit new market opportunities. Senior managers burdened by detail decisions because there was no clear business-management level below them to take responsibility.

- Conglomerates, with a wide variety of different businesses, organized under group vice presidents and "rationalized" so as to share common facilities. Typically these were situations in which newly acquired businesses were digested and assigned to be serviced from common manufacturing plants.
- Manufacturing plants that had evolved to the point where they serviced many different kinds of businesses with varied product lines.
- A major business segment undergoing a complete reorientation of its activities because of significant changes in market conditions. General management felt the need to recalibrate priorities and make sure that the functional organizations understood the implications of the new strategy.
- Manufacturing plants that as a result of a cost-price squeeze needed to sharpen their missions to make them more responsive to the economic pressures in the businesses they serviced.

One might think such conditions would not exist in the 1980s. After all, great effort has gone into the process of empowering separate business units that encourage the "intrapreneurial" manager. In fact, however, these conditions are found in many segments of American industry.

After decades of expansion, diversification, and "rationalization," we now have a complex interplay of business activities struggling for survival inside the typical industrial organization. Most of them are "half pregnant": plants serve more than one profit center; research is centralized; a common sales organization at the regional level probably shares the distribution system and all the support services. This is not called a matrix structure because "independence" and "entrepreneurial management" are much preferred. But in fact it is a matrix that has become increasingly complicated because of the new pressures on interrelationships between the functions, pressures that stem from the need for greater control and faster response. The workshop process provides a mechanism for recognizing these issues and determining how to work them.

THE KINDS OF ISSUES

Four broad kinds of issues have come up during the workshop process:

• Integrating the functions
• Rationalizing the functional and the business missions
• Allocating resources
• Implementing the business strategy

Integrating the Functions

All the functional organizations needed preparatory planning before they could be effectively integrated into a new business strategy. The integration of the manufacturing function often presented the most difficult task by far, because marketing and general business managers had more or less assumed that the manufacturing plants could adapt to whatever could be strategized.

The workshops revealed, with emphasis and detail, that the manufacturing function required time, effort, detailed planning, and in some cases systems and capital support before it could respond to any substantial changes in the business strategy.

Three cases have been selected from the workshop experience that are related to the manufacturing function. The situations are real, though the names given to the companies involved are fictional.

1. Manufacturing management unable at first to respond to a combined plan worked out by the product design engineers and by marketing (Prairie Motors)
2. Manufacturing management unprepared to provide the higher service-level responsiveness required by the marketing program (Kitchen Appliances)
3. Manufacturing overwhelmed by design engineering in the timetable of new-product introduction (Test Equipment Systems)

Prairie Motors

Prairie Motors manufactured a line of specialty electric motors for certain commercial and industrial uses. Its special-market niches and willingness to custom design had generally kept it out of direct competition with high-volume Japanese motors. As a result they had avoided the pressures that had forced other American manufacturers to reduce costs and become competitive or shut down.

Prairie's method of special-market niche was so effective that two once-successful strategies eventually emerged as self-destructive. First, sales and marketing were encouraged to accept orders no matter how small for virtually any motor specifications. Second, capital investment for the plant was held back in the belief that the existing machines and work force were attuned to a variety of products and that the company's market would insulate it from price competition. The approach resulted in significant profits over a long period during which other companies in the industry had gone bankrupt. Then the company was sold to a large corporation for a high price.

Almost as soon as the new management took over, the failings of the previous strategies caught up with the company. A market decline occurred at the same time that specialty motors were introduced from Japan at significantly lower prices and higher quality. It quickly became clear that drastic action had to be taken.

Fortunately the company's product engineers had developed designs along modular lines for a new motor. The modular design could satisfy a significant percentage of the existing specialty market while reducing manufacturing costs and improving reliability. The engineering department had worked out the proposal fully, but since prior management had not authorized it as an action program it had not been closely examined by the other functions. The manufacturing managers, however, had indicated that in general they were in favor of it.

When the modular concept was presented in some detail at the workshop, the marketing people endorsed the proposal. The salespeople felt some concern that the low-volume end of the custom business would be lost, but they welcomed the price reductions and quality improvements that the concept seemed to offer. The manufacturing managers, however, put a "go-slow" on the proposal at this point. The reasons became obvious with a close examination of the modular design. Almost nothing in the existing plant facilities and equipment was ready for it.

- The plant manufactured a large portion of its own machined parts and housings; none of the equipment was cost-effective for long runs of modular parts.
- The assembly area was designed for flexible support of a wide variety of batch runs; there was no significant use of assembly technology and the plant workers were not ready for it.
- The materials support was designed for a wide variety of special parts and the cycle time through the plant was conceived of in terms of months rather than the weeks or days that the new concept called for.
- The shop stewards were not responsive to change and regarded negatively the disciplines that would be imposed by a fast-cycle operation.
- Considerable new capital investment would be required to support the plant.
- The modular design had not been checked for producibility and would have to be reworked with the assembly technology in mind.

The workshop thus served to indicate how much detailed planning was necessary to integrate manufacturing into the engineering and marketing plans before the new program could be effectively implemented.

Kitchen Appliances

The product line of Kitchen Appliances was well regarded in the industry, but the company was coming under increasing price pressure. As part of the company's search for nonprice means of improving value, the marketing department had been asked to explain to the workshop its program to improve the service level of a wide line of small appliances.

The new program required so significant a reduction in lead

time from order to delivery that some types of products would have to be shipped within 48 hours. The implications for the buildup of inventory were staggering. If the manufacturing plant could not change its response time to smaller batch runs, the amount of inventory investment would be prohibitive.

Discussion indicated that the burden on the plant would be a heavy one. The entire product line was assembled on a single flow line, with batch runs of specialty products mixed with long runs of commodity products. There was no way to separate small appliances from the rest of the line, and at the same time the company needed the volume of the large commodity appliances to maintain the plant as a going concern.

The importance of the small appliances became the issue. Salespeople reported that distributors often purchased the commodity products only because the company offered good service on the small appliances; they warned that these promises had to be made good.

Both marketing and sales had committed to the new program and were already making promises for improved delivery to selected customers. The manufacturing managers got them to agree to hold back public announcement until the systems people determined what was needed to support the program. Without the installation of new facilities that would restructure the order system to give priority to the smaller appliance lines, batch runs of the commodity appliances would be interrupted or delayed.

Test Equipment Systems

Test Equipment Systems was a high-technology growth organization that made test equipment for certain kinds of electronic components. It was driven by its brilliant scientist-turned-manager founder who continually brought out new products.

The continual change had always been difficult for the manufacturing people, but they recognized this as the style of the organization and went to great lengths to make do. They worked long hours "picking up after the engineers" and working with immature designs on the shop floor. They went out of their way to anticipate problems by catching the design engineers before new concepts had been frozen and discussing ways to make new designs producible.

This arrangement had brought the company success in the 1970s, but in the 1980s it seemed to cause nothing but difficulty. The market had become more mature and very competitive, test equipment was far more complex than during the previous decade and use requirements less forgiving. Equipment costs had gone up rapidly and customers made very careful comparisons among competing equipment before they purchased. Although Test Equipment Systems products had the best performance in most areas, they were also the most expensive.

The company began to lose orders to competitors, including some Japanese companies, that provided many of the same performance features at lower cost. With these circumstances, the president of the company issued a directive to cut the manufactured cost of the main product by 20 percent.

The manufacturing manager pointed out that two actions would be necessary. First, the engineers would have to undertake a complete redesign of the product to make use of lower-cost components and to enable the utilization of automatic-assembly equipment in the plant. Second, additional capital was needed to obtain the new equipment and to train workers in its use.

The company president was taken aback by the amount of capital the manufacturing manager considered necessary and by the degree of technical support requested. He had become so accustomed to the responsiveness of the manufacturing organization to the continual introduction of superficially new product designs that he had not realized what it would require to undertake a fundamental redesign of the line for cost containment.

Rationalizing the Functional and the Business Missions

The various functional organizations made workshop presentations about their progress along the lines of improvement they had set for themselves. After several of such presentations it began to be obvious that something was wrong. If so much progress was being made, why was the overall performance of the business so poor?

The general discussion periods revealed all too clearly that the progress was simply not in line with the improvements required by the business.

The disarray was partly the result of "turf" issues; the managers improved their own organization in any way possible in the belief that good functional performance was the best base from which to support any business strategy. It was also partly the result of unclear objectives: the plans of the different functions had not been translated into each other's language.

Three cases that relate to rationalizing the mission have been selected.

1. Marketing trying to enter a new technology while manufacturing remains organized to improve its cost-effective support for the old one (Poly-Hexane Corporation)
2. Manufacturing trying to develop a participative organization, while changes in the marketing and technology require greater control over the detailed work disciplines (Gulfhaven)
3. Marketing broadening the product line, causing a cumulative burden on the plant, while the manufacturing managers are trying to improve the effectiveness of existing equipment, which is not configured to make a wide line of products (Saturn Paint)

Poly-Hexane

The Poly-Hexane Corporation, with its innovative product research, spearheaded the technology in the research and manufacture of polyhexane materials and held the leading market position. Its manufacturing plants were designed solely around the production of polyhexane compounds for sealing applications.

Despite the superiority of the company's products, there were competing technologies at the upper end of the scale, in the polyheptanes; at the lower end of the scale, in the polypentanes. Increasing performance requirements in some of the markets traditionally supplied by the Poly-Hexane Corporation began to place the company at a disadvantage in relation to a new generation of polyheptane products; increasing price pressures led to the penetration of the polypentanes in other markets.

Technically it was not difficult for the company to design products in either of these different fields. Small-reactor scale-up tests indicated they could readily be manufactured. Still maintaining its leadership in hexane technology, the company introduced products in both the heptane and the pentane types.

The workshop revealed that the marketing plans were sound but something was wrong with the cost calculations. Usual costs in hexane manufacture were maintained, but the plants were unable to stay within the limits established for heptane or pentane manufacture.

A task force was formed to examine the problem. It found that the existing general-purpose equipment had been so adapted to the needs of hexane technology over the years that the new products' different requirements were slowing down reaction times. The reactors appeared to be general-purpose types, but they had become tuned to hexane cycle times, pressures, feed rates, and mixing systems. In fact, computerizing the reactors was instrumenting them even more to the band width of hexane requirements. The equipment was clumsy and the work skills dysfunctional when pentanes or heptanes were run.

The manufacturing plant was advised to organize two technical teams to adapt operating practices to heptanes and pentanes, and to reduce effort on the hexane program.

Gulfhaven

The Gulfhaven Corporation produced some items with extremely long cycle times. They were difficult to make, required special facilities and considerable skills, and needed the closest possible supervision. A portion of the Gulfhaven plant was dedicated to these products. New equipment was brought in specifically for them and a computerized system was installed to control the process. Yet plant performance did not meet expectations. Batches of product were found to be defective and had to be scrapped; improvement after the introductory phase of the operation was slow; and manufacturing costs were running higher than anticipated.

For the workshop, the manufacturing team was asked to present a detailed analysis of the difficulties along with recommendations or action plans to correct them. Instead, the team emphasized its worker participation program. Under this system, there was little direct supervision; jobs were rotated monthly so that every worker could get to know the entire operation; few disciplines were followed in the manufacturing sequence; the records kept were too few to help identify mistakes or trace their origin. The team believed that this participative philosophy would by itself lead to more effective output by informed workers.

Workshop discussion indicated otherwise. There was a good deal of sentiment in favor of a more disciplined approach. The rest of the organization felt that the plant most needed the detailed work disciplines that would enable it to get the drill correct for the long cycle processes.

A task force was formed to examine the issue. It reported that turnover was high in the plant so that the participative approach was not leading to operator expertise; the computerized control system was not fully automatic and required considerable operator intervention. It had also found that many of the operators themselves would welcome closer supervision and disciplined procedures to make sure that the batches came out right.

There was general agreement that a worker participation program would be a valuable approach—in a plant where the process technology had already matured.

Saturn Paint

Saturn Paint produced a wide line of paints and coatings that were well received in the market. The manufacturing plants were attuned to the requirements, and the overall cost performance was thought to be as good as any in the industry.

During the 1980s, the research group began introducing new formulations to upgrade the product line and keep it ahead of competition. Each new formulation was only a slight departure from the old one and the research and development engineers were able to demonstrate that it would cause only slight stress to existing manufacturing processes.

When the manufacturing people were contacted, however, they

claimed that the cumulative effect of all these slight departures was creating a major burden. Since some of the formulations stressed the equipment in one direction and others stressed it along another set of parameters, both the equipment and the workers were having a difficult time keeping up. Cost performance was deteriorating, cleanouts between batch runs were becoming more frequent, and some batches did not meet performance criteria. In addition, it was at this point that the marketing people said they were planning to introduce a whole range of new products.

The issue that emerged from the workshop was the longer-term pattern of the product mix and its effect on production. When products that had not yet been released for production were added to the existing ones and compared with the product mix that the company had been making, the extent of the future manufacturing problem began to come clear.

Before the increased product line was introduced, a task force was organized to launch a product-rationalization program and to standardize some of the design elements so that further variety would not make the manufacturing task impossible.

Allocation of Resources

The workshops made it plain that the general managers and the marketing managers were often unwilling to face the fact that the programs they were generating might require considerable capital investment in the plants. In some cases shifts in market requirements and the reformulation of strategy had not been thought through far enough to anticipate the effect on the support functions. This was in part because not enough people had been brought into the picture or recognized the significance of the changes. Often, however, it was simply because the homework had not been done to determine what the impact would be.

Seen in full, the capital implications tended to make or break the program. Sometimes the priorities had to be completely shifted because the company was unwilling to support new facilities; other times the new facilities were discovered to have so much potential that they obtained more support than had been originally anticipated.

Three cases that relate to the allocation of resources have been selected.

1. A support division that became an "orphan" in the corporation because none of the customer divisions wanted to provide the funding needed to modernize its plants (Vacuum Division, Central Corporation)
2. A new manufacturing process that presented more potential for future improvement in product technology and cost performance than had been generally recognized (Hermes Process, Kenworth Pharmaceuticals)
3. An apparently mature product sector that actually had significant growth potential provided it could shift its resources to the new products (Cash Cow Division, Central Industries)

Vacuum Division, Central Corporation

The Central Corporation had developed a major program for the United States Navy that involved the production of large and complex electronics countermeasures equipment. Most components and materials were either manufactured by the Systems Division itself or by outside vendors who came under the quality-control programs instituted by the division for bought-out materials. A curious exception to this rule were the inhibitor materials made by the Vacuum Division of the corporation.

The Vacuum Division had a mixed reputation in the Central Corporation. On the one hand, the other divisions that used their materials freely admitted that no equivalent products could be obtained from outside vendors and that their use enabled them to design circuits to tolerances that were competitively superior. On the other hand, they complained that these materials were not under statistical quality control, and that they frequently had to compensate for off-grade quality and delayed shipments. The Vacuum Division workshop began to piece the story together.

Vacuum Division managers agreed with the complaints of the products managers and added a few of their own. They had been so constrained by corporate policy that they had lost the ability to act like an effective business team. They were not permitted to sell their products to outside customers because it might give too much of a technical advantage to competitors. But neither were they given the means to make their products effectively for inside sale. The transfer price was based on a formula that passed the product along at fully loaded cost. However, the manufacturing facilities were so inadequate that they were not capable of meeting the performance standards that were now required.

The equipment had been set up on a hardship basis 20 years earlier. The plan was to outfit the plant properly if and when the technology was successful. The technology had succeeded, but the corporation was facing a recession when it came time to facilitize the plant. From then on everything was continued on a make-do basis. The division had never been refacilitized, and they had no independent market means of retaining funds to upgrade themselves: they were not permitted to sell outside, they were not permitted to increase the transfer price, and they were not authorized the funds needed to modernize. It was a Catch-22, and they had eventually shrugged their shoulders and gone along with the existing unsatisfactory situation.

In this case the situation had to be brought to the attention of the senior corporate officers before any effective action could be taken.

Hermes Process, Kenworth Pharmaceuticals

Kenworth Pharmaceuticals relied on an established manufacturing process to separate impurities from its main product, a new kind of antibiotic, and to refine and size the particles and prepare the materials for activation. A workshop was arranged when the company was considering alternative ways to invest in the modernization and expansion of the plant. Much of the discussion centered on the size and location of the new facility. However, the discussion shifted immediately when a presentation was made regarding the Hermes Process.

The Hermes Process was a new approach to the purification phase of the process that promised to reduce capital investment and provide greater flexibility. The company regarded it as a promising development not yet ready for full-scale operation; although it had been funded on a small scale it was not generally taken seriously nor was it widely reported throughout the company. The workshop presentation revealed that the process was further along than most of the managers in the company had expected; it had been successful in every test run assigned to it, the economics looked favorable, and it was capable of handling a wide variety of new classes of product.

When the marketing and product research people realized the implications of the process, they suggested three or four new product areas and asked for an assessment of the applicability of the Hermes Process. The process engineers and the manufacturing people could not see why it would not apply. In fact, when the capital investment, operating costs, and versatility of the process were compared with the alternative of modernizing the existing process, it became clear that Hermes offered a potential breakthrough for the company. The greatest potential, in fact, appeared to lie in the use of the process for a variety of new products rather than in improvement of the established line.

Once the group faced the size of the capital investment, it began looking for additional product support to pay it off—and discovered that it might have a gold mine on its hands. The sense of the meeting was so positive that task forces were formed to shift resources to the Hermes Process on a scale to service a broader range of products.

Cash Cow Division, Central Industries

The Cash Cow division of Central Industries had long been considered to be operating in such mature markets that keeping it profitable was hardly more than a custodial task. Some new products had been identified by the marketing people and developed by research, but these had not changed senior management's view that the business was mature. Consequently, whenever the manufacturing plants had requested capital investment to carry out a modernization, they had been turned down on the basis that their mission was to generate cash for the corporation, not absorb it.

The workshop revealed that underneath the general pattern of mature products, there was a very dynamic picture of change. The mature product lines were under price pressure on all fronts and some of them were declining. Projections of future sales indicated that further price erosion and decline in unit volume could be expected as foreign competition penetrated more of their market. On the other hand, the growth products were farther along than had generally been supposed, and one of them was moving into high-volume sales. Projections confirmed that the market potential was real, the products superior to those of competitors, and that there was a valid need for them.

The workshop generated a task force to reallocate resources throughout all the functions, away from the old product segments and into the new ones. In particular, it served as an early warning to the manufacturing plants as to the shift in product mix they could expect and the kinds of requirements they would be obliged to meet. It also helped loosen up capital investment for the new growth products.

Implementing the Business Strategy

The workshops in most cases constituted a tuning-up process to bring the functions in line with the implementation of the strategy. From a manufacturing point of view, they served to introduce the managers to the full implications of the business strategies. From a business point of view, they often became an eye-opener that revealed the extent of the support required if the original strategy were pursued without adaptation to the constraints of the manufacturing plants.

Four cases that relate to implementing the business strategy have been selected.

1. The plant not staffed or structured to meet the kind of price competition the company was facing from foreign imports (Capulet Synthetic Sheets)
2. The business that had taken on a stage of forward integration becoming too complex for the plant to support on a cost-effective basis (Canasta Printing Blankets)
3. The business that had to rationalize its own product lines and those of a recent acquisition before any significant progress could be made in manufacturing cost-effectiveness (Trans-American Connectors)

4. The company whose marketing group was planning a shift of product line that the manufacturing function was in no position to support (Specialty Steel)

Capulet Synthetic Sheets

The Capulet Building Products Corporation pioneered in the 1950s the introduction of a certain type of synthetic rubber sheet for commercial construction. The product had superior performance characteristics to justify its higher price, and the company supported it heavily with technical assistance that enabled users to convert their operations. A large-scale manufacturing plant was constructed at a heavy capital investment, and this enabled the company to produce large numbers of sheets at efficient costs. Once the introductory phase was over, the company enjoyed a large volume market and high margins.

By the 1970s, however, the market was becoming mature. Competitors had entered the market and were selling comparable products at somewhat lower prices. Although the Capulet name and the confidence it inspired enabled the company to hold its own volume, larger-volume distributors had become dominant; these were supplied both by Capulet and its competitors as well. The distributors argued that since Capulet's technical support was no longer needed by the customers, their prices were too high.

In the 1980s, the competitive situation became severe as the result of a hard dollar combined with cheap foreign imports from Taiwan and other Asian and South American countries. At first the imports were shrugged off on two grounds: they were aimed only at the commodity segment of the market, and they were lower in quality than the Capulet line. However, imports increased each year and caused a round of pricecutting by Capulet's domestic competitors as well.

As a result of this situation, Capulet management formulated a new strategy. The market would be separated into a commodity segment and a specialty segment. In the commodity segment, the company would service accounts but not attempt to meet the prices of imported products. In the specialty segment, every effort would be made to provide high-quality products and rapid customer service. The sales order administration, the field sales, the inventory management, and the plant production economics would all be primarily oriented to the delivery of superior specialty products at prices competitive enough to hold both this market segment and the traditionally high margins associated with it.

During the workshop, each of the functional organizations presented its plan to carry out this new specialty products mission. It all sounded convincing, with the exception of two key presentations. First, the marketing people reported that the quality of the imported products was improving and that they were entering the specialty products categories. Furthermore, the imports were dropping prices to where they were delivering product at prices comparable to the production costs of the major Capulet plant. Second, the manufacturing people went through their process capabilities in some detail. Several facts became clear.

The plant had been originally designed to make products in high volume. Over the years it had been adapted to a product mix of both commodity and specialty products. However, running the plant only for the specialty products and without the volume of the commodity line would not be economically viable.

There was no known way to make labor costs competitive with foreign factories. The workers in the main plant had grown up with the business; many were 20- and 30-year veterans with all the associated benefits of an established work force. Attrition would not reduce the manning tables quickly enough to meet the current competitive situation, and efficiency improvements would not reduce the work content sufficiently to take up the slack.

At some time during the subsequent decade, the plant would need a major modernization program. Maintenance costs were already increasing, and substantial infusions of new capital would be required to support the operations.

In other words, although the specialty products strategy sounded convincing, conditions in the market were deteriorating more rapidly than expected. The manufacturing plant was shown to be less capable of such a response than had been assumed: neither the facilities nor the work force were in a position to carry out the new strategy if they lost the base in commodity products and faced a new wave of price pressure on the specialty products segment. The most likely solution appeared to be a closure of the plant, while a different kind of flexible manufacturing operation was developed in an area with lower labor costs.

Canasta Printing Blankets

In the late 1970s, Fiber Products of America identified a growth market in the printing-blanket market and determined that their existing vinyl technology was applicable to it. After a market feasibility study with positive results, a research program was authorized to identify the basic elements of a product line. The program defined three specific product bases, which were soon approved as working projects.

By the early 1980s, the three products had been introduced to the market and the company entered a period of growth. The board of directors expected the products to develop into a major business. However, a workshop revealed some negative market conditions.

The products were intended to enable printers to switch from certain pigments under suspicion as possible carcinogens to other nontoxic pigments. The market strategy had been based on the assumption that the EPA would enforce a crackdown on the old pigments that would drive printers to convert to the new blanket. But the EPA decided to continue studying the matter without engaging any strict enforcement program. Meanwhile, suppliers of the suspect inks were working with equipment producers to develop containment procedures that would limit exposure to the suspected carcinogens during the printing process.

The successful three products had to be customized to the specific printing equipment of the end user. Instead of developing three families of products, the research department had developed three specific products, each one of which was different enough from the others as to require different scale-up and manufacturing controls.

The scale-up process appeared to have particular technical difficulties that made it nonlinear to volume increases. As the volume went from samples to small batches to production levels, the nature of the process controls apparently changed in ways not yet fully understood by the technical and manufacturing people in charge. They were confident that they would be able to scale-up each product for manufacture—but not within the present time limitations allowed to them.

In this case, the strategic plan had been convincing to senior management on two grounds: the general agreement that there would be a conversion market driven by environmental pressures, and the assumption that the technology would permit whole families of products to be derived from three basic modules. Instead, there had been no compelling force from the environmental side, and the technology seemed to be moving toward the development of dozens if not hundreds of different product specifications, each one of which would be costly and difficult to scale up for manufacture.

Trans-American Connectors

Trans-American Connector Corporation was a well-established company with a long history of making electrical and electronic connectors. Over the years the product lines had proliferated considerably, as different end markets generated new specifications without withdrawing the old ones. The company supplied connectors for military and commercial aircraft equipment and both industrial and consumer markets. The standard-product catalogue ran to hundreds of pages, and the company also had a make-to-order department to supply special runs of custom designs.

This enormous proliferation put pressure on the manufacturing operations. Most of the equipment had been obtained years before and had been arrayed in a general-purpose machine shop configuration. Batches were released to the shop and then sent to the assembly area. However, with the passing of time the widening line meant smaller batch runs. The manufacturing managers had attempted to cope with the situation by replacing the old machine tools with programmable machine centers. But the large amount of programming effort this required always seemed to be behind schedule. In addition, the assembly workers had to go through a learning curve for every new batch. For these reasons the company had tended to run batches in larger unit volumes than were immediately needed so that there would be fewer runs through the calendar year. Unfortunately, this built up the inventory to such an extent that the senior managers became alarmed at the investment and the possibility of holding obsolete products.

Meanwhile, the parent corporation acquired another major corporation with a division that made connectors in three separate plants. Headquarters people believed that the lines were supplementary rather than competitive. A general manager was therefore placed in charge of the combined business with instructions to rationalize the line and reduce costs.

The workshop revealed a considerable overlap in product lines and manufacturing plants. Four plants were to be rationalized, but each required a clarification of its mission before it could be fitted into some larger plan of action. The home company's original plant was very large and produced the whole range of product. The three newly acquired plants produced overlapping product lines; not one was limited to a specific type and range of activity. It became clear that the general manager could not reduce costs more than marginally until some fundamental decisions were made. The workshop served the purpose of clarifying priorities.

The manufacturing manager suggested some alternative plant rationalization programs. He proposed shutting down two of the new plants and expanding one of the remaining two to specialize in military business and the other to specialize in commercial business. As one alternative, he showed what could be done with a significantly reduced product line, provided that the overall unit volume would not then decline more than 15 percent. A third proposal showed what would have to be done to modernize the existing plants if there were no product line rationalization and no plant shutdowns; it became immediately clear that the company could not afford to modernize all the plants or to support all the products at competitive manufacturing costs.

A consensus emerged that action had to be taken by the marketing people in concert with the manufacturing, materials, and engineering groups. The product line needed to be rationalized before any modernization program could take place.

The new general manager gradually developed a program that

separated the standard commodity products from the specials and the military. The standard line was assigned to one plant and the others to the more expensive but highly skilled labor force. Large numbers of low-volume products were dropped from the line and the plant with the standard products was refacilitized for assembly line operation.

Specialty Steel Company

Specialty Steel produced special steel shapes for the construction industry along the Pacific Coast. Over the years, as the original equipment became more and more out of touch with the product mix, the plant facilities had become stretched. Stamping, bending, casting, plating, and heat-treatment facilities had been configured for a certain product mix. As time went on, however, some of the requirements increased so much that some facilities had become bottlenecks and others were out of use much of the time.

The company had sold some steel to the ship construction industry, but such sales had always been confined to a small portion of the total volume and to shapes similar to those used in the construction industry. However, a market study indicated that an expanded naval construction program presented an opportunity to support that market with special shapes and castings that the shipyards could not readily make up themselves.

The program had been announced at various management meetings, and the general manager assumed that everybody was working toward that end. Unfortunately, the full implications of the new thrust had been entirely missed by the plant. The plant people began complaining about the increased requirements for castings—at just the time the foundry was limited by EPA regulations.

At the workshop, the marketing people prepared a more complete projection of their new market thrust than they had so far. If they were successful, they concluded that within three years the castings requirements would double and the metal bending and treatment requirements would virtually disappear. The presentors kept referring to market projections that had previously been distributed through the organization. These prior documents, however, had been written in terms of market segments, penetration strategies, competitive positions, price levels, and so on. They had not probed the full significance of the program, and they were nowhere translated into terms that the plant could pick up and respond to.

When the presentations ended and the discussion period started, the plant manager got up and said: "Well, if that's what you want, you're not going to get it, because we don't have that kind of plant."

The production people pointed out to marketing that they had frequently raised the issue of the environmental problems and had assumed that everyone understood that it might mean the foundry would have to be closed down. Certainly, the foundry could not be expanded to the level the new program called for without major funding.

The general manager had understood the issue, but he had been trying to deal with it on an incremental basis—a little more castings volume each year and a little less bending. The workshop confronted him with a clear decision as to whether to support the new marketing program or not. In this case, he decided not.

WORKSHOP RESULTS: AN OVERVIEW

There is no question that the workshops' success as a communications vehicle helps the functional managers work as a business team. Largely because of their participative quality, they stirred up much potential in the organizations involved and made these functional managers much more ready and willing to integrate their planning. The managers learned what was and what was not important to the other functions, and at the very least they took the trouble to check with each other on the most critical factors.

Looked at in terms of hard, identifiable business results, the workshops brought about a good deal of steady progress, some real breakthroughs, and a few failures.

When the task force recommendations are compared to the approval rate, the most solid progress was generally made when the power bosses got together to work out the problems among them; they were usually able to shift priorities among themselves and get the results they wanted. When there were too many staffers on the task forces, gains in insight and understanding may have been made, but when the staffers turned in their recommendations they still had the selling job to do. Sometimes the line bosses took action; sometimes they did not.

When the task forces were set to work on new strategic approaches, the results were mixed. In a few cases they came up with innovative results; more often they merely tuned up the main strategic thrust.

The greatest difficulty arose when a task force made a unanimous recommendation for some new kind of resource—a plant modernization program or a new kind of technical support— that they then failed to get. In most of these situations, the battle still goes on; the team may win its point in the end. But it can be frustrating for a management team to get all dressed up with no place to go.

These latter difficulties could fall into two broad categories. First were the situations where the suggestions or actions were the right ones, but the group did not get what it wanted. In one such case, the task forces put together a well-coordinated plan of action to move in one direction, just as a new flood of cheap foreign imports held them up and forced them to move in another. In another, there was general support for a new program all the way up the line until the final review killed off the proposal on the basis of budget limitations. This caused a certain amount of dismay at the middle levels of the organization, and they asked their group vice president for a comment. "Welcome to management," he said.

The second category involved situations in which inadequate guidance led to rejections that could have been avoided. In one case, the group vice president "didn't want to interfere" and "wanted to let it all well up from below," so he let the task forces go wherever they wanted. They came up with a proposal that would have gold-plated the plant, and it was shot down with some enthusiasm. In another case, a task force worked out a new and innovative approach to the market that had great potential—but it crossed the boundaries carved out by another part of the organization. When their group vice president did not support it, it failed to please.

There were some notable successes.

In one case, the workshop and its task forces built up such a weight of opinion against continuing the present strategy, which they felt had long outlived its usefulness, that senior management dropped it and let them develop a new approach.

In another, the workshop brought out such damaging testimony on the obsolete status of one of the plants, and provided such convincing documentation on the new process technology used by competitors, that senior management authorized a major modernization of the plant.

In yet another, such effective evidence was produced as to the disorientation brought about by new-product introductions that one of the timetables was stretched to allow the plant to take action to prepare for it. Nothing catastrophic happened and the company improved its reputation in the market for product reliability.

The workshop process was generally successful in opening up Management Row Game concepts and trade-off methods to test strategies and reinforce their implementation. The results were mixed in situations where middle management wanted something that ran counter to senior management expectations.

This raises the whole problem of the vertical interface between senior corporate management and middle-level business management. It is a difficult and important issue, particularly where the manufacturing function is concerned. It will be discussed in the chapter that follows.

9

Implementing Business Strategy in Industrial Corporations

Trying to implement a new business strategy through an established structure is something like trying to get an elephant to dance—it is possible, but it is seldom graceful. The workshop process has taught that there are ways of making it happen. Four general conclusions have emerged.

1. Strategic thinking needs a thoroughgoing shakedown. Convincing strategic concepts could not be implemented in a number of instances because of limitations in the in-house capabilities of the functions or because of a lack of management understanding of the resources that needed to be deployed.
2. Most of the implementation difficulties come down to a limited number of identifiable stress points in the organization. The same stresses probably show up in most industrial organizations whenever they are pushed beyond their design limits by stretched objectives.
3. Since many of the stress points are identifiable in advance, they can be audited. Moreover, it should not prove impossible to put programs in motion that improve the organization's ability to implement business plans, even if the details of the strategic plan are not as yet entirely defined.

4. Diversification and conglomeration have made the management job more difficult. Procedures have to be established that will sort out the priorities of one business strategy over another when both are obliged to call on a common pool of resources. The trend toward multipurpose manufacturing plants has made some plants unable to do any one thing superlatively well.

It should be kept in mind that two issues commonly dealt with by students and practitioners of the strategic arts have not been emphasized in this discussion of the workshop process.

Consideration of whether a strategy was "good" or "bad" was ruled out of bounds. Whatever strategy had been laid down or was being formulated for the purpose of the workshops was accepted; the sole question was whether or not it could be implemented by the existing resources. Nor were the external factors that influence the success of a strategy judged. Competitive, market, technological, and other "environmental" forces were evaluated; but no judgment was made as to whether the environment was favorable. The answer to one question was sought: Does this business have the resources to carry out its plan?

In many cases so much management energy was centered on the strategy that the necessary resources to carry it out were seldom examined to the depth required. When an appropriate examination was carried out, serious deficiencies invariably appeared.

AUDITING THE FUNCTIONS FOR CAPABILITY OF RESPONSE

When task forces were set in motion to deal with implementation, they were obliged to review the limitations of their functional resources. Four principal limitations appeared in the manufacturing function.

- Constraints of the manufacturing facilities
- Limitations of the labor force
- Weakness of the support services
- Combined activities in the plants

Manufacturing Facilities

The constraints of the manufacturing facilities could be traced to four types of problems:

First, manufacturing processes that featured some special stage that created a bottleneck in the plant—a precision machine that required a long and careful setup, a chemical-process unit, a heat-treating oven, or a plating unit. It can be thought of as a manufacturing stage that required the product-in-being to wait until the process became available, was prepared for it, and then did something extremely slowly that had to be cleaned out later.

Second, machines or process units that were simply not designed to meet the product tolerances they were called upon to provide. They had been tuned up, instrumented, and controlled in every way the industrial engineers could imagine, but they were not capable of meeting the new requirements. After the product came out of these machines they were sorted to keep those that happened to fall within tolerances away from the others. The cost and time involved boggle the mind.

Third, machines that were perfectly adequate but no longer configured to meet the existing product mix or batch size. At some mythical time in the distant past they had been exactly in tune with the business needs; since then the product line had expanded to larger batches for the commodity segment of the market and to smaller batches for the specialty segment. It was not unlike a procrustean bed, with the manufacturing people trying to chop a large run into a series of small batches to get them through the equipment, or running the equipment at a fraction of its capacity to deal with the new slate of low-volume specialty products.

Fourth, problems in plant layout and the movement of materials caused by obstructions and expansions made for good reasons in the past but now blocking the way to rational management. Sequential processes were located at opposite ends of the factory; boiler rooms sat astride the main flow of materials; stock locations so moved around from area to area that they were no longer anywhere near the process units that used them.

Labor Force

The labor force in any plant develops a character of its own. It can be experienced or inexperienced, stable or subject to high turnover, structured or flexible, keyed to long-term or short-term types of remuneration. It can be participative and involved in the improvement of the work tasks or neutral and uninterested in anything but the job as it is strictly defined.

A labor force and a manufacturing function are always in a process of adaptation to each other. Personnel policy may exist to change the work culture or maintain it despite demographic changes taking place in the area. Conversely, there may be little active attention to the work culture from either labor or management.

9-1. Labor and the business mission.

Business Mission	Labor Force
Market dominance	Holding to the pace of repetitive work
Specialty market niche	Care with special product specifications
Delivery response	Willingness to accept changing work schedules
Market coverage response	Willingness to change work assignments
Custom product response	Careful training for equipment changeovers
Product innovation	Participation in developing new work methods
Technical innovation	Innovative ability and technical skills

Whatever the work culture, it has to be reckoned with. It is not something that can be turned on and off, and it has a critical role to play in the manufacturing mission and the implementation of strategy.

Going back to the Row Game, figure 9-1 suggests that the labor force associated with a strategy of market dominance would be very different from one that would best support a strategy of innovation. The mass-production requirements of market dominance mean that labor has to accept the pressures associated with a paced production line. Innovation needs labor to be flexible, respond to changing work assignments, actively contribute ideas, and develop expertise in the job's technical requirements. This does not mean "involvement" is impossible among mass-production workers through quality circles and flexible job assignments; rather that the inherent limitations of the mass-production plant are different from those with an innovative mission.

Personnel policy and the manufacturing function, however, can get very much out of sync with each other. The personnel function may attract people who are marching to a different drummer than that of the plant foreman. Unless personnel managers reckon with the plant mission they can use personnel as a test bed for a variety of social experiments.

What is to be done, for example, when problems like these arise:

- A participative work force has been created and it is then discovered that in order to survive the plant has to be mechanized.
- A "disciplined" work force is waiting to be told what to do—and active participation is needed to help shake down the new products.
- Labor peace has been bought with successive pay increases and then the company finds that it has priced itself out of the market.

Support Services

The manufacturing support services—particularly the materials management staff and the industrial engineers—are becoming a critical issue in many companies. They are "overhead" because their work is not directly allocatable to specific products; yet they may be as important to the plant's direct output as the line workers. And in many companies it is the expense of overhead, not direct labor, that has grown to dominate the manufacturing costs.

If the plant mission requires flexibility and the managers have not thought through the implications or the needed funding has been turned down, what generally tends to happen? The managers build up a force of indirect staff, materials people and methods engineers, expeditors and staff assistants, all of them trying to turn the elephant into a ballet dancer. The investment needs are covered up by the slow growth of overhead charges.

That is one way to get the job done. But it may turn out to be cheaper in the long run to get to the heart of the process technology and design it for the task it is required to do.

Combined Activities

The effect of combining activities is to make more difficult every one of the problems we have identified thus far. What can be done about this situation?

It is true that the problem is inherent in the complexity of modern business, but not all that can be done is being tended to. Chapter 2 suggested a four-stage approach.

1. Rationalize the overall plant array by clarifying the manufacturing needs of the different businesses.
2. Develop a principal mission for each multibusiness plant.
3. Establish factories-within-the-factory to take care of all the businesses.
4. Set up a management review system to look after the needs of the smaller businesses.

This places a good deal of pressure on the management review process to make sure that the system works. It is an inherently complex task that is seldom managed as carefully as it should be.

INTERFUNCTIONAL DIFFICULTIES

When the workshop task forces began to come to terms with the manufacturing limitations, they found themselves dealing with several interfunctional difficulties. At first glance these appeared unique to the business or the plant; but defined properly, they came down to a limited set of problem areas that showed up—in quite a variety of clever disguises—in virtually every workshop.

In the manufacturing area, five problems emerged again and again.

The accounting system: Unfortunately, the accounting system seldom sheds light on the underlying nature of these costs. For strategic analysis, it is necessary to know the very kinds of things about overhead that the cost system cannot explain.

- How much of the overhead cost is associated with a strategy of frequent new-product introduction?
- How much is associated with product proliferation?
- How much with old plant or old production methods, which could be eliminated by modernization?
- How much with the task of making an inherently inflexible manufacturing facility more responsive to change?

If it is correct that a good deal of the overhead is traceable to the cost of variety, new ways of measuring it are needed. W. L. Wallace's paper in this volume's appendix suggests a framework related to the concept of the "hidden factory" described by J. G. Miller and T. E. Vollmann, among others. The number of "transactions" caused by changes to the manufacturing task are measured to correlate the cost of the support staff with the complexity of the task it is trying to achieve.

The workshops made clear that the manufacturing managers knew a number of ways to reduce manufacturing costs. But they all required some support from the other functions to limit the product slate, to redesign the problem products, to hold to agreed-upon schedules or provide more lead time, or to provide some capital funding for new equipment. Once they got past the impasse caused by the cost system's failure to show the way, it was possible to get manufacturing and the other functional managers together to work out an approach.

Meeting product performance requirements: Sales and marketing people are concerned because the products are not meeting the best competitive standards, they are receiving complaints from customers, or else products are not meeting the in-house test requirements.

One would suppose this to be a strictly technical problem. Some performance parameter in the product must be improved and some particular mechanism in the factory will have to be replaced.

In fact, in almost every case clarification through an interfunctional management group was required before the problem could be turned over to the technicians. Once defined clearly enough to be passed to the technicians, the problem was usually capable of fairly ready resolution.

Why did something specific like product tolerances require the consideration of an interfunctional management group?

- There were different views as to what performance parameters were required to meet the market strategy.
- There were different approaches to obtaining the performance.
- The performance factors they were most concerned about could not be measured directly, and the plant was testing other parameters that related to them.
- Conditions in use were very different from the conditions of test.
- The sales and marketing people had a hierarchy of performance requirements, some of them considered essential and others merely desirable.
- The design engineers developed their own hierarchy of performance requirements that did not always correspond to the ones of the marketing people.
- There was a cost factor that changed the priority once it moved too far above or below some undefined target level.

The product specifications passed on to the technicians consisted of several feasible alternatives, depending on the cost and time required to find the fix. This occasionally required a second task force made up of design and industrial engineers, purchasing and manufacturing people, in order to lay down the best line of approach.

What did general management learn—or not learn, in some cases—from this exercise?

- Any change in the required product performance is going to generate a massive amount of detail at the middle levels of the organization.
- There is a limit to the number of specifications changes that can be handled by the organization.
- It is a responsibility of general management to determine how much effort to apply to this sort of task and what the priorities are.

Changing unit-volume requirements: Customers across all of industry have discovered that it is better not to hold inventory themselves if they can get their suppliers to hold it. That is why just-in-time has become popular. But if it is to work properly, the system requires close communication between the supply plant and the customer plant. In many cases this requirement emerged without the requisite communications system. For example, the communications link does not exist when a product is moved through outside distributors; consequently the plant is turned on and off without warning.

Again, an interfunctional task has to be defined before the technicians can take over. Why did an art as arcane as inventory management require managers from sales, marketing, purchasing, and manufacturing to deliberate the issue before the systems people could take over?

- Nobody could define a service-level objective in a form usable by the technicians without coming to terms with the requirements of different customers and different products.
- There was a close relationship between the cost of the service level and a willingness to provide it.

- There were some economic conditions when the required course of action was obvious and others when it was not at all clear.
- Production schedules interact with each other, and improvements in one area can cause delays in others.
- The amount of detail required to solve the problem was so large that it had to be prioritized at the management level so that it would not get out of hand.

Changing product mix: The problem of product mix is consistently underestimated by nonmanufacturing people. Time and again, marketing people state something to the effect that they had projected demand in advance and that they were only off by, say, 10 or 15 percent. So what are the manufacturing people complaining about?

The manufacturing people are complaining that the product slate they are required to make is different from the one they had been expecting. They cannot issue schedules until they determine the specific products that have to be made, and when these are changed at the last minute—sometimes during a run—the whole plant is disrupted.

Once again, why was there a management-level understanding to be reached before the specific problems of product changeover could be addressed?

- There were different kinds of understanding as to what constituted a lead time.
- Some changes in some products or processes required little advance notice while other kinds generated the greatest possible confusion.
- Delays in the communications links caused the utmost resentment—for example, some people in home office knew there would be changes in the schedule and failed to notify the plant on a red alert basis.
- The nature of the changes was not thought out in advance and prepared for, and instead was imposed on the system as one more crisis.

Unfortunately the cost pressures are usually generated when the business gets into some kind of crisis. Orders come down to cut everything back and reduce the support staff. Since the support staff is often the only means by which any kind of orderly cost reduction can take place, this is something like taking a soldier's rifle away from him and telling him to fight harder.

The timetable: Most timetable problems are associated with new-product introduction, but these are not the only cases. It is more correct to say that new-product introductions invariably place stress on the plant timetable and that other situations do so as well—notably, the changes in product performance requirements, in production schedules, and in product mix that have been noted above.

The manufacturing function contains an enormous amount of detail to be mastered, because it organizes work for large numbers of people and integrates their work with the cycle times of the machines. With the introduction of a significant change, all of the calculations have to be rebalanced. The plant can do this, but it is best done within the disciplines that have been organized for it. Stress is created by the imposition of timetables generated outside the system that do not take orderly procedures into account.

Events in a process industry illustrate such stress. The order came down to reduce inventories, and it was done. When the demand picked up again, the entire orderly sequence of batch lot production had been thrown into disarray. The plant was off balance, had to interrupt long runs in order to meet emergency customer needs, and had to change its regular sequence of optimized cleanouts to meet the short-term requirements of the product slate.

Here as before defining acceptable timetables required interfunctional management discussion. Why was it not a simple thing to clarify?

- Certain stages of the procedure could be changed quickly, while others required very considerable preparation.
- There were technical problems that had to be resolved.
- Some machines might have to be altered or replaced.

- The response depended in part on work commitments else-where in the plant.

THE CLASSIC BUSINESS PROBLEMS

All functional problems identified so far are traceable to stresses imposed on the organization by a new strategy or the more vig-orous prosecution of an existing one. The stresses are themselves traceable to a limited number of classic business problems. Here are the most prominent ones:

Product Proliferation

Product proliferation occurs when a new product is introduced to the line and an existing one is not taken out, yet the overall unit volume does not increase commensurately. It may be a re-sponse to the natural evolution of the product life cycle, in which more and more special needs and niches are exposed. The problem does not lie in the fact that proliferation may be bad - for the business; it may be very good for the business. The benefit is easy to identify; the problem lies in the fact that the benefit's cost is difficult to determine.

In the past consultants consistently pointed out proliferation's costs—shorter batch runs, more changeovers, loss of learning curve, more exposure to quality failures, increased inventory, burdensome bookkeeping, exposure to errors in order admin-istration, and so on. Today's business managers understand the problem—they just do not do much about it. A stronger con-stituency is usually in favor of making every potential customer happy than a constituency for rationalizing manufacturing and inventory. But the cost pressures remain unresolved, and that enormous, amorphous burden of manufacturing overheads is back again. Why does it require an interfunctional task force to resolve this?

- The manufacturing people are the only ones in a position to understand the alternative ways of manufacturing product but they lack cost details.
- Accountants can project the costs of minor variations around the existing product slate, but they are not in a position to

know how much the costs could be reduced if the plant were given the option of manufacturing for a significantly narrower product slate.

- Salespeople see the benefit of a wide product line, but they are seldom given the opportunity to weigh it against the benefit of a narrow product line in cost reduction and delivery responsiveness.
- The general business manager seldom has the luxury of seeing the problem presented in the form of two alternative game plans, complete with sales-volume projections, manufacturing strategies, and product-cost data.

New-product Introduction

The key to most strategies lies in a campaign centered around a new product or technology; early emphasis in planning lies in the research and development and the marketing areas. Cost projections are made, which involve the plant; and market coverage is assessed, which involves the sales function—but by that time the strategy is pretty well crystallized. For the strategy to be implemented, however, the whole formidable apparatus of the transition from engineering to production has to operate effectively. This difficult area reaches well beyond the small group of industrial or manufacturing engineers in the middle. It involves the nature of the product design, its maturity, and the timetable of its completion, as well as the manufacturing organization's equipment and work skills.

The process of transition, however, has become an increasingly identifiable activity. It can be audited and made ready for a new marketing campaign. It is not absolutely necessary to wait until everything has gone wrong before studying the problem. The disciplines outlined in chapter 6 provide a valid checklist for this task; management overcommitment can be documented by assessing the new program's risks.

Customizing the Product Line

As we mentioned above, the natural evolution of the product life cycle generally moves one block of the product slate toward increasing the customizing of the product line to meet the special needs of certain market niches. In addition, we are witnessing a general shift in competitive product offerings toward the needs of different customers.

In today's industrial markets, more customers than before want just-in-time delivery, statistically controlled quality, and special designs or formulations to meet their specifications. Some of the larger customers want exclusive rights to a particular design. Components feeding into military and government products have to meet special requirements; materials for medical and food products have to meet others. In the consumer markets, the range of product offerings in virtually every field has become greatly extended, while at the same time their market life has shortened. So the general effect is a long way past the story about the bad old days of the Model T: any color as long as it is black. The other side of that Ford story, incidentally, is that it was priced retail at something under 500 dollars.

"Commoditization" of the Product Line

Along with increased customizing of one part of the product line has come a rapid growth in the commoditization of the other. The Japanese have been particularly deft in this dangerous game. The strategy of this kind of penetration involves anticipating market dominance, establishing the manufacturing plant and the distribution channels to support it, and then flooding the market with product priced significantly below the competition to make the kill.

This dangerous game does not always work and tends to destroy the market; in many cases the net result is simply to break market prices for everyone and share the same volume at lower margins. But it is a game that has to be played with the right cards or not at all—and this means a tremendous pressure on the manufacturing organization to meet stringent cost targets.

Certain difficulties surfaced through the workshops.

In numerous cases the business was not the market aggressor and was trying to fight a defensive battle against a new wave of price competition. Unless the marketing people panicked and cut prices across the board, the first line of defense was always to improve the special customer services and try to differentiate the product to keep it competitive. This is fine as long as it works, but experience indicated that management generally realized too late that whatever else happened, manufacturing costs had to come down—no matter what ugly implications this may have had for labor negotiations, plant location, and capital funding.

In many cases management did not separate the specialty from the commodity products and thus got the worst of both worlds. This was especially true of the plant operations, where everything remained mixed together, thoroughly blocking any effort to get a firm grip on the costs of the commodity part of the line.

The Sales-Marketing Interface

A new marketing strategy has to be implemented through the field sales function. However, at the close of the day's business the salespeople have to report the orders that they booked; there is a limit to how much missionary work they are able to support, let alone their technical competence to support it. They like nothing better than to talk about new products and introduce them, but their role as market intelligence does not work very well (nobody seems to read their customer reports). They have trouble getting samples out of the plant and the customers always remind them of the last new product they introduced that failed. The marketing people, meanwhile, do not seem to know when or how to let go. They often ignore their own salespeople in order to find out the same things from contract market research firms; they like to make missionary calls themselves. At any rate, it is a troubled area, severely stressed by any change in strategy.

The Engineering-Marketing Interface

It is not an easy thing to get a good fix on the future needs of customer. They do not always know themselves ("Would you buy an Edsel?" they were asked, and the answer was: "Why, certainly!"). Industrial end-users may be blocked from their vendors by an impenetrable battery of reprocessors and product assemblers or distributors who want to keep that end-user relationship to themselves.

In addition, there is now more technical content in most products than there used to be. It often happens that marketing people have to proceed along one line of advance to anticipate and define the end use requirements, while the technical people proceed along an entirely different line of advance in order to further the state of the art. Marketing will lose every time it tries to dictate terms to the research community: you can't bully Mother Nature, and you can't push research people around to the point where they might consider migration to a competitor. Research is going to lose just as heavily when it tries to dictate terms to the marketing community: you can't bully the customers into buying some brilliant technical achievement that they can't use or can't afford. So it is an uneasy relationship, and a new strategy can make it more tense than ever. When the relationship works, however, there is a natural sympathy and mutual respect between the two groups; if the strategy cannot build on that rather than place stress on it, there will be trouble.

THE SPECIAL DIFFICULTIES OF THE MANUFACTURING FUNCTION

Clearly, there are some areas where the poor orphans who become manufacturing managers might be generally agreed to have become the aggrieved parties in industrial organizations:

The Fault of the Strategists

When new-product technology requires a new plant, the strategists generally provide the manufacturing function with the best resources.

However, where the technology is going to be fitted into an existing plant, the recklessness is staggering. People seem to think that whatever it is, the plant can make do and get on with it.

There seem to be two reasons for this kind of neglect, which is particularly evident in the case of the multipurpose plant that is obliged to service different business requirements.

First, the strategic business plan is drafted in marketing terms and financial terms, but seldom in manufacturing terms. The senior people will tolerate minor amounts of funding support but they do not want to hear that anything major will be involved, because that is likely to sink the program.

Second, once a plant has been written down, its depreciated value makes for a nifty-looking return on investment for a mature product line. Even the people from marketing or sales who complain about poor quality and slow response are reluctant to see the asset base increased because at least initially that puts more pressure on them to maintain margins and increase sales.

Finding a Champion to Support the Capital Funding

"Manufacturing people are always trying to get more capital funding out of the company; but they can't fool us, we're not going to give them any." So say many managers.

Demands for capital investment are always great, if a company has let them go for a while they can be overwhelming. Furthermore, they never stop. A manufacturing plant is like your grandmother's house: once you finally get to clean out the cellar, the roof starts leaking.

The problem is to find a corporate champion to support capital funding when none of the strategic business units cares to step forward. It is difficult, because general modernization programs lack appeal. There are two reasons for this.

First, they look so postponable. If the profit center managers are all trying to achieve certain objectives when the plant manager submits a general-purpose program for improved stock control, it is difficult for any senior manager to place the appropriation request very high on the list of immediate priorities. The request may be approved if there is extra cash flow that quarter, but delaying the request creates no real problem.

Second, the traditional calculation for return on investment can be very misleading. The criteria applied to plant modernization programs assume that these manufacturing activities will reduce product costs. In fact, however, the costs may be closely related to the marketing programs of one or more of the businesses that the plant serves. The base case in these cases should not be the present cost structure but rather the cost structure that would obtain over the life of the investment if the modernization program is not authorized. Looked at in this way, the base-case scenario might well project lower product volume and a deteriorating situation in the market. What if the plant is unable to improve its delivery service and product quality, and runs into difficulties supporting an increasingly varied product line? How will these conditions affect the marketing programs of the businesses?

A champion from the marketing side of the house is needed, one who can understand these issues and fight for the kind of manufacturing plant that will help him spearhead his own programs.

Difficulties Related to Manufacturing Operations Generally Require Broad Support to Be Resolved

It was something of a surprise to us to realize how much of the manufacturing activity was in fact a ballet involving the other functions.

Time and again manufacturing people will be pressed for action on some point; they will work it through to where it comes to a standstill and waits for resolution through marketing, sales, or engineering. Specifications need to be redefined, service levels authorized, or technical uncertainties pinned down, which manufacturing is unable to do on its own.

Many of these requirements impose a new burden on the technical staff at the service of the manufacturing people—manufacturing engineering or those who define specifications for incoming materials. However, this staff is generally considered plant overhead and is therefore on the suspect list.

GENERAL CONCLUSIONS ABOUT THE IMPLEMENTATION OF STRATEGY

One underlying problem seems to come up again and again. It is related to management decision-making and business planning in the complex industrial corporations constructed in recent years. If effective implementation of business strategy is to be ensured, it is clear that a lot of detail has to be worked among the various functions at the middle levels of the organization.

Yet decisions must be made at the top of the corporation with little backup detail and no first-hand knowledge of the current capabilities of the various functional organizations. This problem in vertical communications shows up in two ways:

- The imposition of "stretch" objectives
- The changing tasks of key managers in a corporate industrial organization

The Imposition of "Stretch" Objectives

This can be described as the difference between slow metabolism and hot desire.

The higher up in the corporation, the more rapidly triggered the expectations become. Senior people are drivers. They can generally taste the sweet results before a program is even off the drawing board, and they speak a conceptual language different from the operating language spoken in the manufacturing plants.

The lower down in the corporation, the more detail exists. Particularly at the plant, any concept has to be shaken down and translated into shop operations that tell each one of a hundred or a thousand different workers what to do when he or she comes to work that day. There is a limit to what such a work organization can absorb. That may be why some manufacturing managers have developed a protective style that buffers their plants against objectives imposed from outside.

By definition, a "stretch" objective is at the upper limit of possibility. Unless it is defined very carefully, it is easy to slip into the land of make-believe. This is not a serious failing if the objectives are psychologically reinforcing; no harm is done and possibly some good. However, in numerous situations the stretch objectives lead to contractual agreements that commit the manufacturing plant to producing an impossible output. The objectives then are not only no longer psychologically reinforcing, they are downright destructive.

It may be asked why stretch objectives are defined here as "imposed." Are they not volunteered from the bottom up by the general business manager? Often that is the case. But a scenario like the one that follows can also create the problem.

In the heady atmosphere of the planning review a kind of pot-latch takes place. The general business manager comes in, billed as the great leader. He (or she) lays down objectives that may possibly be achieved provided nothing goes wrong anywhere down the line inside the organization and the competitors remain docile outside it. The senior managers ask if he has thought of this and thought of that, and he assures them that everything is under control. Then the party is scheduled and the general business manager, playing host, prepares to throw all his worldly goods into the festivities.

It is only later in the sober light of afterthought that he begins to wonder why he insisted on committing his organization to accomplishing so many outlandish things.

It would seem that such stretch situations can be identified easily enough. What is needed is the courage to monitor them and call for a review early enough in the implementation to do something about any problems.

Changing Tasks of Corporate Managers

The roles of the key players in an industrial organization need careful thought in today's changing industrial world.

Corporate Headquarters

The tasks facing senior managers in corporate headquarters have changed during the 1970s and 1980s, but comparable changes have not always been made in the roles they carry out.

The changes in tasks have been considerable. Businesses have grown and become more complex. Acquisitions increase enormously the variety of markets and types of businesses that are carried on. The rate of change within each market is higher than before because of extension of competition to an international arena and to the explosive effects of technological change. The sheer weight of management decision-making and control has increased greatly at the top of the industrial corporation.

Senior managers have at the same time become more and more involved in the new pressures exerted on the corporation from outside. They are bound to be involved in litigation of one kind or another because it is an inescapable fact of life these days, in the shape of product-liability suits, tax cases, or antitrust issues. The corporation is probably involved in an acquisition or else it is defending itself from an outside move. Meanwhile the financial community is watching its every action and interpreting the anticipated results favorably or unfavorably.

Despite all this change, some corporate managements are still heavily involved in the operational decision-making processes. They want to probe into the details whenever they feel uneasy about something. They may hold up each appropriation request until it can be matched and weighed against every other one.

In the early 1960s, Harold Geneen, then president of the ITT Corporation, described what he felt about the delegation of management decision-making in a modern international industrial corporation. He did not believe in it. He wanted to be part of every significant decision in the corporation, and he went to great lengths to absorb enough detail to attempt it. He said: "If I had enough time, I would do it all myself." But that system fell of its own weight when Geneen retired, and nobody has volunteered to revive it elsewhere on a comparable scale. The fact is that even Harold Geneen could not cope with the detail facing a corporation of the 1980s.

Changing Tasks of Group Managers

The tasks facing group general managers seem also to have evolved without comparable changes in their roles. Many corporate group general managers have as large a job and as complex a variety of businesses to operate as the chief operating officer of a corporation did a decade ago. The businesses have grown and changed and new ones have been acquired. In an attempt to simplify the structure, the tendency has been to divide things up and assign them to a limited number of group managers—everything in each group made coherent only by that elusive concept called "synergy."

Yet it is often true that in the eyes of corporate management the group managers may constitute little more than a convenience. In such organizations their role comes down to a process of digestion: they tidy up their business so that corporate management can then pass judgment on them.

With an eye to the complexity of the 1990s, it is difficult to understand how corporations will be able to function if group managers acquire more responsibilities without being given commensurate authority to make decisions.

Some evidence of this unresolved problem appeared in the workshops. A business manager would obtain general agreement on a plan of action, and then find that to get the staffing and capital support, he or she would have to wait until another level of management above the group level reviewed the case.

The best arrangement of tasks and roles would be a power structure in which the group managers have fewer businesses reporting to them; in which they can reliably tell the managers reporting to them what support to expect so that the strategic business managers can get on with their jobs and take action when they need to.

Implications for General Business Managers

For good or bad, the action has got to lie with the middle level general managers of the strategic business units. Everything else is little more than prioritization, coordination, and support. These are the people who must call the shots and deploy their businesses in competition.

If that is the case, then the business managers must play the Management Row Game in earnest, despite the difficulties of working within a large corporation. They must behave as if they owned all their resources whether they do or not.

This has great consequences for the manufacturing function when the plants service many different businesses. The general business manager must understand the manufacturing resources thoroughly so that they can be integrated into the game plan. The implications for the plant of future changes in the product mix and the marketing requirements must be understood, and the product technology and the process technology must be in tune with each other.

The general business manager must *not* allow himself to become a marketing manager. It is he or she who must integrate the activities of the manufacturing, engineering, and sales functions.

Implications for the Manufacturing Managers

I think it means a great deal. Manufacturing managers will have to think increasingly in business terms if they are to understand what they can contribute to the competitive capability of each of the businesses they serve. The way is open to a new breed, one that will work on a much broader range than the traditional plant manager of the 1970s.

He will see his plant conceptually not just as the place where "the product gets out the door." He will think in terms of systems and materials flows, not just in terms of operations and direct costs.

He will know what the best manufacturing practices in his industry consist of, making the rounds outside his own shop to see what can be done with his kind of process anywhere in the world.

He will be able to translate the marketing demands and the product mix requirements of the strategic business units into manufacturing terms of flow requirements and process control— not just at the plant, but in discussion with the other managers of the businesses he supports.

He will know where the strategic business unit is heading and what kind of plant it will need in the future, so that every stage in modernization works toward an objective rather than merely toward solving an immediate problem.

He will understand the trade-offs and how to work actively with other functional managers reporting to the general business manager, so as to be able to reach agreements in the common interest.

He will know what to fight for and what to let go.

In other words, he—or she—will become a modern corporate business manager, not just a traditional production person. When the job is seen to be as demanding and as significant as all this suggests, more of the best people will continue to go into it. They will stop hiding behind the image of "I'm just a tough little old production man, so you tell me what you want and I'll try to get it for you."

They will learn how to thrive inside a modern industrial corporation.

Appendix

A Different Way of Thinking about Manufacturing Costs

William L. Wallace

There are inherent and natural limitations to the information generated by cost accounting systems. This is not to say that there is necessarily anything wrong with our cost accounting systems and data, but rather to point out and emphasize what they don't and can't do.

We should remember that cost accounting systems generally are designed to serve three purposes: (1) to provide a legally acceptable basis for valuing inventory and cost of goods sold for tax and reporting purposes; (2) to provide the basis for direct labor variance reporting, and usually material variance reporting; and (3) to provide some insight into the differences in standard cost between various products within a plant. Cost accounting is usually based on the arbitrary and generally unreal assumption that overhead costs of a burden center can reasonably be spread or allocated on the basis of standard direct labor costs. This is a practical, simple, and consistent way of dealing with burden application, but obviously not "accurate" in many particular cases. Two common examples will readily illustrate this commonplace fact:

1. Product #1 within the plant is processed through highly automated equipment while product #2 is largely done with

hand work with minimal tooling, equipment, and fixturing. In most cases, the normal cost accounting practice of allocating burden by direct labor will significantly understate the "real" cost of product #1 and overstate the "real" cost of product #2.

2. Product #3 is normally produced in relatively long runs with infrequent setup and changeover, while product #4 is normally run in very small quantities, each lot requiring setups at many steps in the process, and requiring extensive material planning, purchasing, production control, and followup action. Both products have equivalent standard direct labor and material.

Real Characteristics

Product #1	Product #2
Low direct labor	High direct labor
High depreciation	Very low depreciation
High maintenance	Almost no mainte-nance
High engineering	Low engineering
High setup and changeover costs	Low setup costs
Low handling costs	Higher handling costs
Low inspection cost	Higher inspection cost

Real Characteristics

Product #3	Product #4
(No direct labor differences)	
Lower purchased parts cost	Higher purchased parts cost
Low setup and changeover	High setup and changeover
Relatively low planning, scheduling, purchasing, and handling costs	Substantially higher planning, purchasing, and handling costs
Low troubleshooting and followup per unit	High troubleshooting and followup per unit
Lower inspection costs	Higher inspection costs

In most cases, our normal cost accounting practice will overstate the "real" cost of product #3 and understate the "real" cost of product #4.

Of course there are other kinds of obvious and inherent limitations to our cost accounting systems with respect to analyzing possible changes in approach. They cannot readily assess the effects of broad changes in process approaches, product line simplification, shifts in scheduling policy, and so on. Also, they do not naturally differentiate between the costs of past commitments, which are still being written off, and the costs which are presently changeable.

None of the foregoing comments should be thought of as criticisms of our cost accounting systems. Rather, they are merely descriptive of the inherent limitations on how data from cost accounting can fruitfully be used in pricing, product line profitability analysis, and strategic manufacturing decision making. For these purposes, we need to have alternative *analytical* approaches which will help provide more meaningful guidance.

Let's consider a possible different way of thinking about manufacturing cost. We can choose to analyze all of our actual and potential manufacturing costs under five categories:

1. The *inherent cost* of manufacturing one of our typical products, as designed, at the volume level of our total throughput for this general category of product.
2. The *cost of variety*, which arises from the fact that we don't make a single product but rather produce a number of variations to achieve the specified breadth of product line.
3. The *cost of change*, which includes the cost of new and changed products, engineering changes, and the cost of scheduling changes, to meet unpredictable or rapidly increasing demand, or to deal with substantial seasonable or cyclical variations in volume.
4. The *cost of system and organizational suboptimization*, which can arise from a number of factors, such as suboptimized inbound and outbound freight, over- or undermanning, over- or undersizing of our facilities, internal frictions within the organization, and so on.

5. The costs of *writing off past decisions* are worth segregating for analysis and decision-making purposes, because they deal only with when and how to expense costs which have already been incurred.

In all five of these categories of cost, we need to deal with both expenses which go through to profit and loss, and with the cost in assets, because capital efficiency has as important an economic effect as current cost.

For each one of these five analytical categories, let us now briefly discuss possible approaches to analysis and some of the ways in which these types of costs may be managed.

Inherent Cost

If there are many varieties in our product line, then it is likely that our present processes and plant layout and organization have been developed overtime with that variety in mind. If so, then to find out what the cost would be if our entire volume were in a single product, we need to make a rough reprocessing, relayout, reorganizing study. If we were to replan from scratch on the basis of steady rate production of a single one of our products, how would we do it and what would it cost in total cost? We obviously would have to revisit all of the expense budgets on a zero-base basis in order to complete this estimate. Typically, this *inherent cost* is 20–70 percent of our present manufacturing cost.

To change the *inherent cost*, we must change the (1) product design or specification, (2) the process, or (3) the volume. To change the product design we might use a value analysis/value engineering approach to be sure that our product is optimized to real customer needs and wants, and that we have developed the design for the lowest cost using known producing technologies. Often designs are based on obsolescent process assumptions because the product designers are not aware of changing process economics. We can change processes either by undertaking R & D programs to develop new processes or by adapting new state-of-the-art processes available elsewhere in the world. Simplification, uninterrupted throughput, and minimum handling should be sought. We can change volume either by riding a market growth curve or by picking up market share by offering more desirable products at lower prices.

Cost of Variety

To analyze the cost of variety in a given product line and plant, one must first analyze both the variety of finished or end items and the number and kind of differing components, subassemblies and materials from which they are produced. In addition, attention must be given to where extra costs arise because of the variety. Some of these extra costs are self-evident, such as setup, startup, scheduling, inventories, and the higher cost of buying in small lot quantities. Other costs are really the cost of missed opportunities, because we cannot afford the best layout, process, and systematization because we must handle lots of variety. Training for multiple tasks and the cost of failure because of multiple tasks are parts of this cost.

To reduce the cost of variety we can reduce the varieties offered to our customers and the unheeded costs of supplying variety. To cut back the varieties in our product line, we can rationalize out of product market segments that are unprofitable, we can rationalize away from customers who are making expensive demands without corresponding prices, we can lead our customers to standardize to fewer varieties by offering price and delivery advantages, and we can routinely prune the items of declining significance.

As important as controlling the actual varieties of products offered is the management of the manner in which the variety is achieved. This means first of all heavy continuing management attention to parts interchangeability, modularity, and design standardization. When differing sizes and features are needed in end products, every effort must be made to limit the degree of difference in purchased and manufactured parts and subassemblies.

It is also possible to change significantly the *cost of variety* in manufacturing by optimizing the way the factory is layed out and managed, so that scheduling and control is both highly effective and not unduly costly. One of the major costs of variety is the loss of productivity that comes from not having the needed parts and materials where and when they are required. Another important control factor is to reduce the cost of setup and change-over by engineering, investment in more flexible equipment and tooling, total concentration of management and supervision, and full involvement of the people. Breaking in to production schedules to run desk orders must also be controlled.

Typically, the *cost of variety* ranges from 20–50 percent of manufactured cost.

The Cost of Change

This cost arises from changes in product, major changes in process, and substantial changes in schedule. It includes all the costs of startup including training and retraining and poor quality. It, of course, includes all the costs of engineering, planning, and implementing the changes. And it includes the costs of balances out and obsolescence in parts, finished goods, equipment, and plant that arise from the phase out of discontinued production. *The cost of change* can be managed by either reducing the amount of change or by increasing the effectiveness with which the organization deals with change.

Typically, the *cost of change* may fall in the 10–40 percent range of total manufactured cost.

The Costs of System and Organizational Suboptimization

This is the cost that derives from internal inefficiencies, rigidities, and lack of communication within the organization of the business, or from freight or communication inefficiencies caused by location.

It is the hardest of the five costs to analyze objectively, both because there is a strong tendency to take the existing situation for granted, and because any questioning of the effectiveness and efficiency of the organization creates tensions and defensive reactions. However, over- and undermanning usually exist. Suboptimal interdepartmental communication and cooperation is commonplace, and few organizations tap all of the motivation and skill of their employees.

This cost can be changed or managed by breaking down artificial barriers to communication and effective cooperation, and by actively enlisting the full contribution that each person in the organization is capable of. Rigorous repeated system and functional analysis is also required.

Typically, the costs of organizational suboptimization could fall in the range of 5–20 percent of manufactured cost.

The Costs of Writing Off Past Decisions

The principal reason for segregating these costs is to prevent confusion in analyzing the financial impact of future courses of action. Idle plant, unfunded pension liabilities, depreciation on equipment and machines that are no longer needed, and the costs of obsolete inventory are all examples of the kind of cost we are talking about.

They are real costs and they will come through to the books of the business and the corporation as losses, either by early write-off or by changes to profit over a period of years. We cannot avoid the eventual impact on our P & L either by writing them off immediately or by taking them overtime. They are not *controllable*.

However, they have nothing to do with the "real" cost of our present operation or of any future alternatives. Hence, we must remove them from the financial analysis of decision considerations, or we will be confusing the usefulness of our analysis.

Bibliography

Abernathy, William J., and Kenneth Wayne. "Limits of the Learning Curve." *Harvard Business Review* 52 (Sep-Oct 1974): 109–19.

Bain, David. *The Productivity Prescription.* New York: McGraw-Hill, 1982.

Buffa, Elwood S. *Meeting the Competitive Challenge.* Homewood, Ill.: Dow-Jones Irwin, 1984.

Hayes, Robert H., and Stephen C. Wheelwright. "The Dynamics of Process-Product Life Cycles." *Harvard Business Review* 57 (Mar-Apr 1979): 127–36.

———. "Link Manufacturing Process and Product Life Cycles." *Harvard Business Review* 57 (Jan-Feb 1979): 133–40.

———. *Restoring Our Competitive Edge: Competing Through Manufacturing.* New York: John Wiley and Sons, 1984.

Hirschmann, Winifred B. "Profit from the Learning Curve." *Harvard Business Review* 42 (Jan-Feb 1964): 125–39.

Malpas, Robert. "The Plant After Next." *Harvard Business Review* 61 (July-Aug 1983): 122–30.

McDonald, Alonzo L. "Of Floating Factories and Mating Dinosaurs." *Harvard Business Review* 64 (Nov-Dec 1986): 82–86.

Miller, Stanley S. "Make Your Plant Manager's Job Manageable." *Harvard Business Review* 61 (Jan-Feb 1983): 68–74.

Porter, Michael E. *Competitive Strategy.* New York: The Free Press, 1980.

Schmenner, Roger W. "Every Factory Has a Life Cycle." *Harvard Business Review* 61 (Mar-Apr 1983): 121–29.

————. *Making Business Location Decisions.* Englewood Cliffs, N.J.: Prentice-Hall, 1982.

Schonberger, Richard J. *Japanese Manufacturing Techniques.* New York: The Free Press, 1982.

Skinner, Wickham. "The Focused Factory." *Harvard Business Review* 52 (May-June 1974): 113–21.

————. *Manufacturing in the Corporate Strategy.* New York: John Wiley and Sons, 1978.

————. *Manufacturing: The Formidable Competitive Weapon.* New York: John Wiley and Sons, 1985.

————. "The Productivity Paradox." *Harvard Business Review* 64 (July-Aug 1986): 55–59.

Wheelwright, Stephen C. "Japan—Where Operations Really Are Strategic." *Harvard Business Review* 59 (July-Aug 1981): 67–74.

Wheelwright, Stephen C., and Robert H. Hayes. "Competing through Manufacturing." *Harvard Business Review* 63 (Jan-Feb 1985): 99–109.

Willoughby, Willis J. Jr. et al. *Solving the Risk Equation in Transitioning from Development to Production.* Washington, D.C.: Defense Science Board Task Force, 1985.

Index